Construction technology

Volume 2

Second edition

R. Chudley M.C.I.O.B.

Chartered Builder

Illustrated by the author

Longman
Scientific &
Technical

Longman Group Limited,
Longman House, Burnt Mill, Harlow,
Essex CM20 2JE, England
and Associated Companies throughout the world

First published 1974
Eighth impression 1985
Second edition 1987
Seventh impression 1996

British Library Cataloguing in Publication Data
Chudley, R.
    Construction technology.—2nd ed.
    Vol. 2
    1. Building
    I. Title
    690   TH145

ISBN 0-582-42037-7

General Editor
C. R. Bassett, B.Sc.
*Formerly Principal Lecturer in the Department of Building and
Surveying, Guildford County Technical College*

Set in IBM Journal 10 on 12 point

Produced by Longman Singapore Publishers (Pte) Ltd.
Printed in Singapore.

Preface and Acknowledgements

The aim of this second volume is to provide a continuity of study with the contents of Volume I, which dealt mainly with domestic building. The second year of a typical construction technology course will study further aspects of domestic construction and introduce the student to framing techniques and materials. The presentation is in the form of notes accompanied by ample illustrations with as little repetition between notes and drawings as possible. It is also written with the assumption that the reader has covered, understood and retained the technical knowledge contained in a basic first year course.

The depth of study presented in this volume is limited to the essential basic knowledge required for a second year construction technology course. The student is therefore urged to consult all other possible sources of reference to obtain a full and thorough understanding of the subject of construction technology.

We are grateful to the following for permission to reproduce copyright material:

British Standards Institution for reference to British Standards Codes of Practice; Building Research Station for extracts from *Building Research Station Digests*; Her Majesty's Stationery Office for extracts from Acts, Regulations and Statutory Instruments.

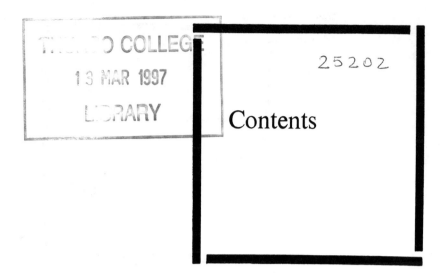

Contents

1
Accommodation, storage and security

Part I
Site and temporary works

The activities and the temporary nature of a building site do not generally justify the provision of permanent buildings for staff accommodation or for the storage of materials. It is, however, within the builder's interest to provide the best facilities which are economically possible for any particular contract; this should promote good relationships between management and staff, it should also reduce the loss of materials due to theft, accidental damage and vandalism. The better the facilities and amenities provided on a building site the greater will be the contentment of the site staff which will ultimately lead to higher productivity.

ACCOMMODATION

The Construction (Health and Welfare) Regulations 1966 is a statutory instrument which sets out the legal requirements for the minimum accommodation and facilities for site staff to be provided on sites throughout the construction industry having regard to the number of employees on site and in some cases the anticipated duration of the contract. The main requirements of this document are shown in Table I.1.

Units of staff accommodation usually come in one of two forms:

1. Sectional timber huts.
2. Mobile caravans or cabins.

No. of persons employed by contractor on site	0 5|10 20 25 40 50	100

FIRST AID
Box to be clearly marked and in charge of named person — First-Aid boxes. → First-Aid boxes and person trained in First Aid.

STRETCHER AMBULANCE
Stretcher provided. Local Health Authority informed of site, work and completion date. If no 'phone or radio, ambulance kept ready.

FIRST-AID ROOM
To be used only for treatment and in charge of trained person. — Where number of persons on site exceeds 250 each employer of more than 40 persons must provide First-Aid room.

SHELTER AND CLOTHING
All persons to have shelter and place for depositing clothing. — Adequate means of warming themselves and drying wet clothing. / Where possible, means of warming themselves and drying wet clothing.

MEALS ROOM
All persons to have drinking water provided and facilities for boiling water and eating meals. — Facilities for heating food if hot meals are not available on site.

WASHING FACILITIES
All persons on site for more than 4 hours to have washing facilities. — Where work is likely to last 6 weeks h. and c. or warm water, soap and towel provided. — Where work is likely to last 12 months 4 wash places plus 1 for every 35 persons more than 100.

SANITARY FACILITIES
To be maintained and kept clean—provision to be made for lighting. — 1 convenience for every 25 persons. — 1 convenience for every 35 persons.

NOTES: Washing facilities to be close to meals room;
Protective clothing to be provided where person is required to work in inclement weather.
Sub-contractors may use the facilities provided by another contractor and for the purpose of these regulations their work force on site is included in the total work force on site.

THE CONSTRUCTION (HEALTH AND WELFARE) REGULATIONS 1966

TABLE 1.1

2

Sectional timber huts are prefabricated to allow for ease of dismantling and assembly to facilitate the re-use on other sites. Huts of this nature should be designed, constructed and maintained with the same care as permanent buildings to ensure their use for many years on a number of different contracts. A well-designed sectional hut should permit the addition of more bays to increase the modular size by length and/or width.

The anticipated use of each hut will govern the construction and facilities required. Offices need to be weatherproof, heated, insulated to conserve the heat, some form of artificial lighting, equipped with furniture such as desks, work tops, plan chests and chairs to suit the office activities; a typical timber sectional site office detail is shown in Fig. I.1. The same basic construction can be used for all other units of accommodation such as meals rooms and toilets equipped with the facilities shown in Table I.1.

Caravans and mobile cabins are available in a wide variety of sizes, styles and applications. The construction is generally of a ply clad timber frame suitably insulated and decorated; they are made to a modular system so that by using special connection units any reasonable plan size and shape is possible. The caravans and cabins are fully equipped with all the necessary furniture, lights and heating units. The toilets are supplied with all the necessary sanitary fittings and plumbing which can be connected to site services or be self contained. Transportation of caravans or cabins can be on any suitable vehicle; caravans can be towed whereas special transporter trailers are available for cabins. Whichever method is used the time taken to load, offload and position on site is considerably shorter than the time required to dismantle, transport and reassemble a sectional timber hut, but the initial capital outlay is higher. Typical examples of units of accommodation are shown in Fig. I.2.

STORAGE

The type of storage facilities required of any particular material will depend upon the following factors:

1. Durability — will it need protection from the elements?
2. Vulnerability to damage.
3. Vulnerability to theft.

Cement, plaster and lime supplied in bag form require a dry store free from draughts which can bring in moist air and may cause an air set of material. These materials should not be stored for long periods on site,

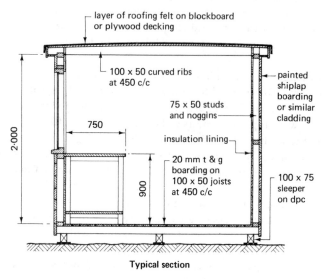

layer of roofing felt on blockboard or plywood decking

100 x 50 curved ribs at 450 c/c

painted shiplap boarding or similar cladding

75 x 50 studs and noggins

insulation lining

2·000

750

20 mm t & g boarding on 100 x 50 joists at 450 c/c

900

100 x 75 sleeper on dpc

Typical section

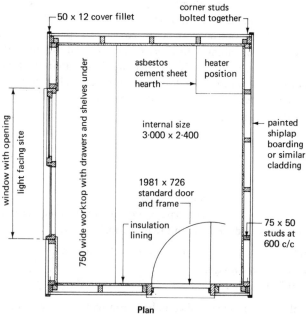

50 x 12 cover fillet

corner studs bolted together

asbestos cement sheet hearth

heater position

window with opening

light facing site

750 wide worktop with drawers and shelves under

internal size 3·000 x 2·400

painted shiplap boarding or similar cladding

1981 x 726 standard door and frame

insulation lining

75 x 50 studs at 600 c/c

Plan

Fig. I.1 Typical prefabricated timber site office

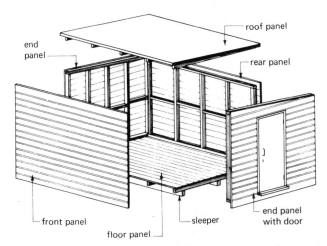

Sectional store hut

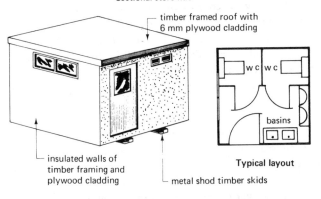

Typical layout

Cabin toilet unit

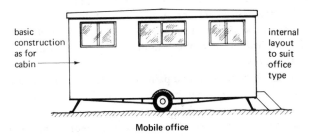

Mobile office

Fig. I.2 Typical site accommodation units

therefore provision should be made for rotational use so that the material being used comes from the older stock.

Aggregates such as sand and ballast require a clean firm base to ensure that foreign matter is not included when extracting materials from the base of the stock pile. Different materials and grades must be kept separated so that the ultimate mix batches are consistent in quality and texture. Care must be taken, by careful supervision, to ensure that the stock piles are not used as a rubbish tip. If the storage piles are exposed to the elements a careful watch should be kept on the moisture content; if this rises it must be allowed to drain after heavy rain or alternatively the water/cement ratio of the mix can be adjusted.

Bricks and blocks should be stacked in stable piles on a level and well-drained surface in a position where double handling is reduced to a minimum. Facing bricks and light-coloured bricks can become discoloured by atmospheric pollution and/or adverse weather conditions; in these situations the brick stacks should be covered with tarpaulin or polythene sheeting adequately secured to prevent dislodgement. Blocks, being less dense than bricks, should be stacked to allow air movement around them and should always be covered with a suitable sheet material.

Roof tiles have a greater resistance to load when it is imposed on the edge; for this reason tiles should be stacked on edge and in pairs, head to tail, to give protection to the nibs. An ideal tile stack would be five to seven rows high with end tiles laid flat to provide an abutment. Tile fittings such as ridge and hip tiles should be kept separate and if possible placed on end.

Drainage goods, like tiles, may be stored in an open compound; they should be stacked with their barrels horizontal and laid with spigots and sockets alternately reversed or placed in layers with the spigots and sockets reversed in alternate layers. Fittings should be kept separate and those like gullies, which can hold water, should be placed upside down.

Timber is a hygroscopic material and therefore to prevent undue moisture movement it should be stored in such a manner that its moisture content remains fairly constant. A rack of scaffold tubulars with a sheet roof covering makes an ideal timber store. The various section sizes allow good air flow around the timber and the roof provides protection from the rain and snow.

Ironmongery, hand tools and paint are some of the most vulnerable items on a building site. Small items such as locks, power drills and cans of paint should be kept in a locked hut and only issued against an authorised stores requisition. Large items like baths can be kept in the compound and suitably protected; it is also good practice only to issue materials from the compound against a requisition order.

SECURITY AND PROTECTION
Fencing

A building site and the compound can be given a degree of protection by surrounding with a fence. The fence fulfils two functions:

1. Defines the limit of the site or compound.
2. Acts as a deterrent to the would-be trespasser or thief.

A fence can be constructed to provide a physical barrier of solid construction or a visual barrier of open work construction. If the site is to be fenced as part of the contract it may be advantageous to carry out this work at the beginning of the site operations. The type of fencing chosen will depend upon the degree of security required, cost implications, type of neighbourhood and duration of contract.

A security fence around the site or compound should be at least 1.800 m high above the ground and include the minimum number of access points which should have a lockable barrier or gate. Standard fences are made in accordance with the recommendations of BS 1722 which covers ten forms of fencing giving suitable methods for both visual and physical barriers; typical examples are shown in Fig. I.3.

Hoardings

These are close boarded fences or barriers erected adjacent to a highway or public footpath to prevent unauthorised persons obtaining access to the site and to provide a degree of protection for the public from the dust and noise associated with building operations. Under Sections 147 and 148 of the Highways Act 1959 it is necessary to obtain written permission from the Local Authority to erect a hoarding. The permission, which is in the form of a licence, sets out the conditions and gives details of duration, provision of footway for the public and the need for lighting during the hours of darkness.

Two forms of hoarding are in common use:

1. Vertical hoardings.
2. Fan hoardings.

The vertical hoardings consist of a series of closed boarded panels securely fixed to resist wind loads and accidental impact loads. It can be free standing or fixed by stays to the external walls of an existing building (see Fig. I.4).

Regulation 46 of the Construction (General Provisions) Regulations 1961 requires protection to be given to persons from falling objects. A fan hoarding fulfils this function by being placed at a level above the normal traffic height and arranged in such a manner that any falling débris is directed back towards the building or scaffold (see Fig. I.4).

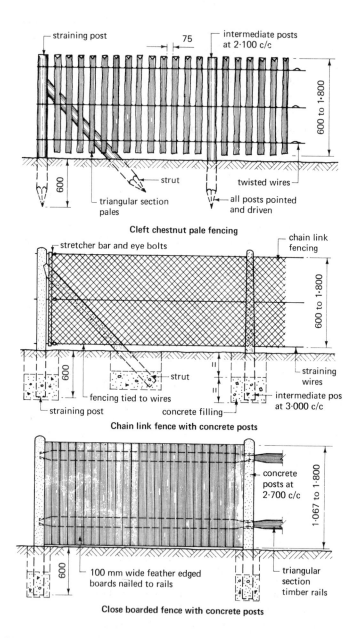

Cleft chestnut pale fencing

Chain link fence with concrete posts

Close boarded fence with concrete posts

Fig. I.3 Typical fencing details

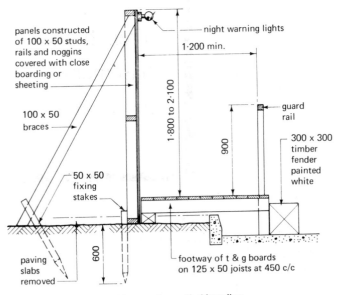

panels constructed
of 100 x 50 studs,
rails and noggins
covered with close
boarding or
sheeting

100 x 50
braces

50 x 50
fixing
stakes

night warning lights

1·200 min.

1·800 to 2·100

900

guard
rail

300 x 300
timber
fender
painted
white

paving
slabs
removed

600

footway of t & g boards
on 125 x 50 joists at 450 c/c

Typical freestanding vertical hoarding

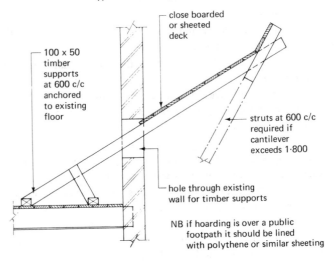

100 x 50
timber
supports
at 600 c/c
anchored
to existing
floor

close boarded
or sheeted
deck

struts at 600 c/c
required if
cantilever
exceeds 1·800

hole through existing
wall for timber supports

NB if hoarding is over a public
footpath it should be lined
with polythene or similar sheeting

Typical fan hoarding

Fig. I.4 Timber hoardings

2

Trench and basement excavations

Excavations may be classified as shallow, medium or deep in the following manner:

1. Shallow — up to 1.500 m deep.
2. Medium — 1.500 to 3.000 m deep.
3. Deep — over 3.000 m deep.

A second year course of study is mainly concerned with the medium depth excavations.

The method of excavation and timbering to be used in any particular case will depend upon a number of factors:

1. Nature of the subsoil can determine the type of plant or hand tools required and the amount of timbering necessary.
2. Purpose of the excavation can determine minimum widths, minimum depths and the placing of support members to give a reasonable working space within the excavation.
3. Presence of ground water may necessitate the need for interlocking timbering, sump pits and pumps; large quantities of ground water may prompt the use of dewatering techniques.
4. Position of the excavation may impose certain restrictions such as the need for a licence or wayleave, Highway Authority or police requirements when excavating in a public road.
5. Non-availability of the right type of plant for bulk excavation may mean a different method must be used.
6. Presence of a large number of services may restrict the use of machinery to such an extent that it becomes uneconomic.
7. The disposal of the excavated spoil may restrict the choice of plant

due to the load and unload cycle not keeping pace with the machine output.

SAFETY

The Construction (General Provisions) Regulations 1961 is a Statutory Instrument which sets out the minimum requirements for the safe conduct of work in excavations for building operations and works of engineering construction.

This document states that an adequate supply of timber or other suitable material must be supplied and used to prevent danger to any person employed in an excavation over 1.200 m deep from a fall or dislodgement of materials forming the sides of an excavation.

The timbering should be carried out as the work proceeds and must be executed under the direction of a competent person who must ensure that all the material used is of adequate strength and suitable for its intended function. All excavations over 1.200 m deep in which persons are employed must be inspected by a competent person at least once a day and excavations over 2.000 m must be inspected before each shift commences. Unworked excavations must have been inspected within the preceding seven days before persons can recommence working within the excavation. Inspection of excavations must be carried out if there has been substantial damage of supports or if there has been an unexpected fall of earth or other material.

A suitable fence or barrier must be provided to the sides of excavations over 2.000 m deep or alternatively they must be securely covered; methods of providing a suitable barrier are shown in Fig. I.5. Materials must not be placed near the edge of any excavation, nor must plant be placed or moved near excavations so that persons working in the excavation are endangered.

TRENCH EXCAVATIONS

Long narrow trenches in firm soil may be excavated to the full depth by mechanical excavators enabling the support timbering to be placed in one continuous operation. Weak and waterlogged ground must be supported before excavation commences by driving timber runners or steel trench sheeting to a position below the formation level or by a drive and dig procedure. In the latter method the runners can be driven to a reasonable depth of approximately 1.500 m followed by an

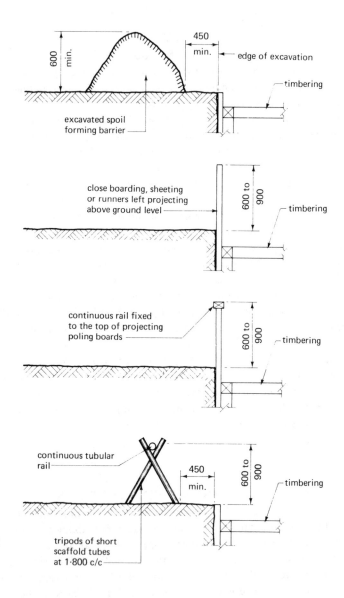

Fig. I.5 Barriers to excavations

12

excavation cut of 1.200 m and then the operation repeated until the required level has been reached; this will make the driving of the runners easier and enable a smaller driving appliance to be used.

In medium depth trenches different soil conditions are very often encountered throughout the depth of the excavation and therefore the method of timbering must be changed to suit the new soil conditions; a typical example of trench timbering in these circumstances is shown in Fig. I.6.

Hand trimming is usually required in the trench bottom to form an accurate line and level; this process is called bottoming of trenches. Approximately 150 mm should be allowed for trimming by hand and it is advisable to cover the trimmed surface with hardcore to protect the soil at formation level from being disturbed or drying out and shrinking.

BASEMENT EXCAVATIONS

There are three methods which can be used for excavating a large pit or basement:

1. Complete excavation with sloping sides.
2. Complete excavation with timbered sides.
3. Perimeter trench method.

Excavation for a basement on an open site can be carried out by cutting the perimeter back to the natural angle of repose of the soil. This method requires sufficient site space around the intended structure for the over excavation. No timbering is required but the savings on the temporary support work must pay for the over excavation and consequent increase in volume of backfilling to be an economic method.

In firm soils where poling boards can be placed after excavation an economic method is to excavate the bulk of the pit and then trim the perimeter, placing the poling boards with their raking struts in position as the work proceeds. Alternatively the base slab could be cast before the perimeter trimming takes place and the rakers anchored to its edge or side (see Fig. I.7).

The perimeter trench method is used where weak soils are encountered; a trench wide enough to enable the retaining walls to be constructed is excavated around the perimeter of the site and timbered according to the soil conditions. The permanent retaining walls are constructed within the trench excavation and the timbering removed, the dumpling or middle can then be excavated and the base cast and joined to the retaining walls (see Fig. I.7). This method could also be used in firm soils when the mechanical excavators required for bulk excavation are not available.

13

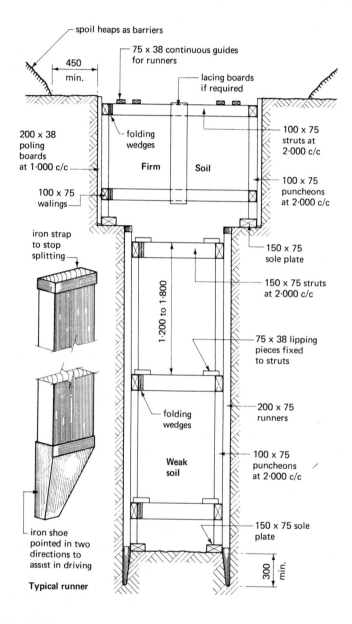

spoil heaps as barriers

75 x 38 continuous guides for runners

450 min.

lacing boards if required

200 x 38 poling boards at 1·000 c/c

folding wedges

Firm   **Soil**

100 x 75 struts at 2·000 c/c

100 x 75 walings

100 x 75 puncheons at 2·000 c/c

iron strap to stop splitting

150 x 75 sole plate

150 x 75 struts at 2·000 c/c

1·200 to 1·800

75 x 38 lipping pieces fixed to struts

folding wedges

200 x 75 runners

Weak soil

100 x 75 puncheons at 2·000 c/c

iron shoe pointed in two directions to assist in driving

150 x 75 sole plate

300 min.

Typical runner

Fig. I.6 Trench timbering in different soils

14

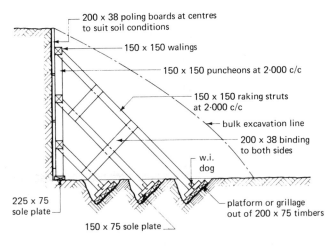

200 x 38 poling boards at centres to suit soil conditions

150 x 150 walings

150 x 150 puncheons at 2·000 c/c

150 x 150 raking struts at 2·000 c/c

bulk excavation line

200 x 38 binding to both sides

w.i. dog

225 x 75 sole plate

platform or grillage out of 200 x 75 timbers

150 x 75 sole plate

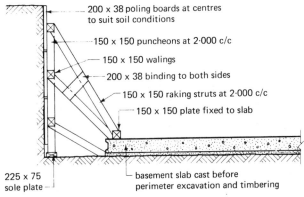

200 x 38 poling boards at centres to suit soil conditions

150 x 150 puncheons at 2·000 c/c

150 x 150 walings

200 x 38 binding to both sides

150 x 150 raking struts at 2·000 c/c

150 x 150 plate fixed to slab

225 x 75 sole plate

basement slab cast before perimeter excavation and timbering

Alternative timbering methods for complete excavation

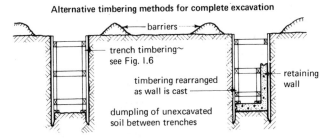

barriers

trench timbering~ see Fig. I.6

timbering rearranged as wall is cast

retaining wall

dumpling of unexcavated soil between trenches

Perimeter trench method

Fig. I.7 Basement excavations and timbering

15

3
Shoring

All forms of shoring are temporary supports applied to a building or structure to comply with the requirements of Regulation 50 of the Construction (General Provisions) Regulations 1961. This regulation requires that all practical precautions shall be taken to avoid danger to any person from collapse of structure. Common situations where shoring may be required are:

1. To give support to walls which are dangerous or are likely to become unstable due to subsidence, bulging or leaning.
2. To avoid failure of sound walls caused by the removal of subjacent support such as where a basement is being constructed near to a sound wall.
3. During demolition works to give support to an adjacent building or structure.
4. To support the upper part of wall during formation of a large opening in the lower section of the wall.
5. To give support to a floor or roof to enable a support wall to be removed and replaced by a beam.

Structural softwood is the usual material used for shoring members; its strength to weight ratio compares favourably with that of structural steel and its adaptability is superior to steel. Shoring arrangements can also be formed by coupling together groups of scaffold tubulars.

SHORING SYSTEMS
There are three basic shoring systems, namely:

1. Dead shoring.
2. Raking shoring.
3. Flying shoring.

Each shoring system has its own function to perform and is based upon the principles of a perfectly symmetrical situation. In practice many shoring problems occur where it is necessary to use combinations of shoring systems and/or unsymmetric arrangements (see Fig. I.8).

Dead shoring

This type of shoring is used to support dead loads which act vertically downwards. In its simplest form it consists of a vertical prop or shore leg with a head plate, sole plate and some means of adjustment for tightening and easing the shore. The usual arrangement is to use two shore legs connected over their heads by a horizontal beam or needle. The loads are transferred by the needle to the shore legs and hence down to a solid bearing surface. It may be necessary to remove pavings and cut holes in suspended timber floors to reach a suitable bearing surface; if a basement is encountered a third horizontal member called a transom will be necessary since it is impracticable to manhandle a shore leg through two storeys. A typical example of this situation is shown in Fig. I.9.

The sequence of operations necessary for a successful dead shoring arrangement can be enumerated thus:

1. Carry out a thorough site investigation to determine:
   (a) number of shores required by ascertaining possible loadings and window positions;
   (b) bearing capacity of soil and floors;
   (c) location of underground services which may have to be avoided or bridged;
2. Fix ceiling struts between suitable head and sole plates to relieve the wall of floor and roof loads. The struts should be positioned as close to the wall as practicable.
3. Strut all window openings within the vicinity of the shores to prevent movement or distortion of the opening. The usual method is to place timber plates against the external reveals and strut between them; in some cases it may be necessary to remove the window frame to provide sufficient bearing surface for the plates.
4. Cut holes through the wall slightly larger in size than the needles.
5. Cut holes through ceilings and floors for the shore legs.
6. Position and level sleepers on a firm base, removing pavings if necessary.

17

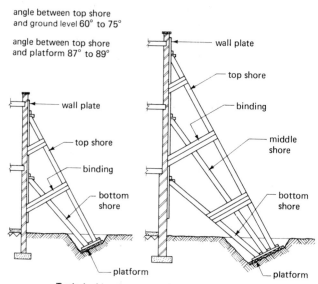

angle between top shore
and ground level 60° to 75°

angle between top shore
and platform 87° to 89°

wall plate

top shore

wall plate

binding

top shore

middle
shore

binding

bottom
shore

bottom
shore

platform

platform

Typical raking shore arrangements (see also Fig. I.11)

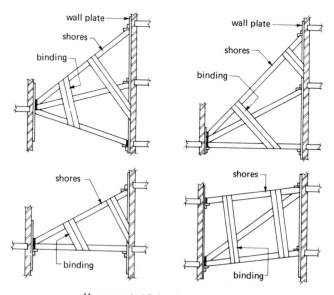

wall plate

shores

binding

wall plate

shores

binding

shores

binding

shores

binding

Unsymmetrical flying shore arrangements

Fig. I.8 Shoring arrangements

18

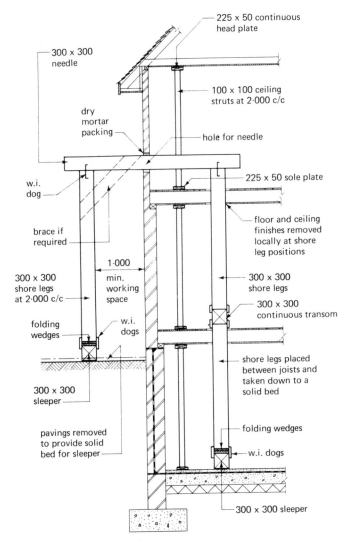

225 x 50 continuous
head plate

300 x 300
needle

100 x 100 ceiling
struts at 2·000 c/c

dry
mortar
packing

hole for needle

225 x 50 sole plate

w.i.
dog

floor and ceiling
finishes removed
locally at shore
leg positions

brace if
required

1·000
min.
working
space

300 x 300
shore legs

300 x 300
shore legs
at 2·000 c/c

300 x 300
continuous transom

folding
wedges

w.i.
dogs

shore legs placed
between joists and
taken down to a
solid bed

300 x 300
sleeper

folding wedges

pavings removed
to provide solid
bed for sleeper

w.i. dogs

300 x 300 sleeper

cross bracing, longitudinal bracing and hoardings
to be fixed as necessary

Fig. I.9 Dead shoring

19

7. Erect, wedge and secure shoring arrangements.

Upon completion of the builder's work it is advisable to leave the shoring in position for at least seven days before easing the supports to ensure the new work has gained sufficient strength to be self supporting.

Raking shores

This shoring arrangement transfers the floor and wall loads to the ground by means of sloping struts or rakers. It is very important that the rakers are positioned correctly so that they are capable of receiving maximum wall and floor loads. The centre line of the raker should intersect with the centre lines of the wall or floor bearing; common situations are detailed in Fig. I.10. One raker for each floor is required and ideally should be at an angle of between 40° and 70° with the horizontal; therefore the number of rakers which can be used is generally limited to three. A four-storey building can be shored by this method if an extra member, called a rider, is added (see Fig. I.11).

The operational sequence for erecting raking shoring can be enumerated thus:

1. Carry out site investigation as described for dead shoring.
2. Mark out and cut mortices and housings in wall plate.
3. Set out and cut holes for needles in external wall.
4. Excavate to a firm bearing subsoil and lay grillage platform and sole plate.
5. Cut and erect rakers commencing with the bottom shore. A notch is cut in the heel so that a crow bar can be used to lever the raker down the sole plate and thus tighten the shore. The angle between sole plate and shores should be at its maximum about 89° to ensure that the tangent point is never reached and not so acute that levering is impracticable.
6. Fix cleats, distance blocks, binding and if necessary cross bracing over the backs of the shores.

Flying shores

These shores fulfil the same functions as a raking shore but have the advantage of providing a clear working space under the shoring. They can be used between any parallel wall surfaces providing the span is not in excess of 12.000 m when the arrangement would become uneconomic. Short spans up to 9.000 m usually have a single horizontal member whereas the larger spans require two horizontal shores to keep the section sizes within the timber range commercially available (see Fig. I.12 and Fig. I.13).

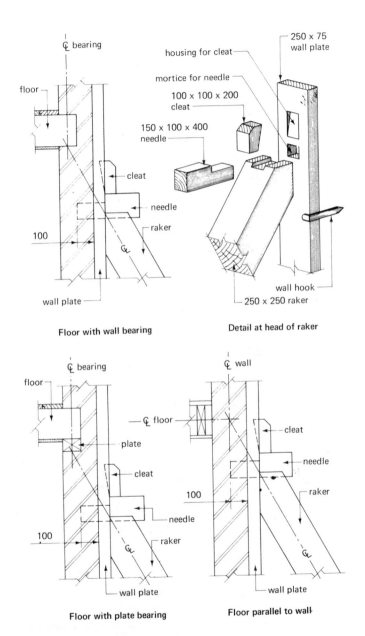

Floor with wall bearing   **Detail at head of raker**

Floor with plate bearing   **Floor parallel to wall**

Fig. I.10 Raking shore intersections

21

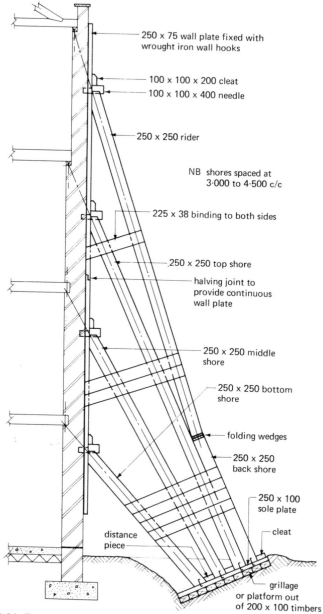

250 x 75 wall plate fixed with wrought iron wall hooks

100 x 100 x 200 cleat
100 x 100 x 400 needle

250 x 250 rider

NB shores spaced at 3·000 to 4·500 c/c

225 x 38 binding to both sides

250 x 250 top shore

halving joint to provide continuous wall plate

250 x 250 middle shore

250 x 250 bottom shore

folding wedges

250 x 250 back shore

250 x 100 sole plate

cleat

distance piece

grillage or platform out of 200 x 100 timbers

Fig. I.11 Typical multiple raking shore

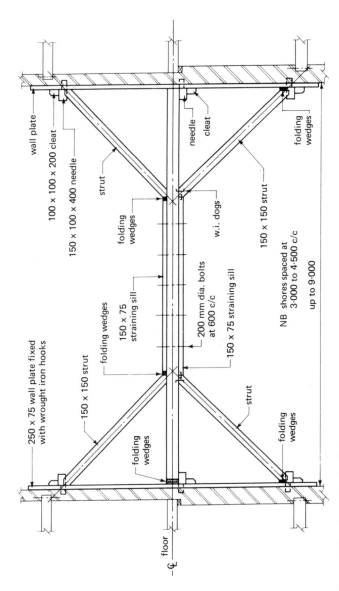

wall plate

100 × 100 × 200 cleat

150 × 100 × 400 needle

strut

needle

cleat

folding wedges

w.i. dogs

150 × 150 strut

folding wedges

150 × 75 straining sill

200 mm dia. bolts at 600 c/c

150 × 75 straining sill

NB shores spaced at 3·000 to 4·500 c/c up to 9·000

250 × 75 wall plate fixed with wrought iron hooks

150 × 150 strut

folding wedges

strut

folding wedges

folding wedges

floor

Fig. I.12 Typical single flying shore

23

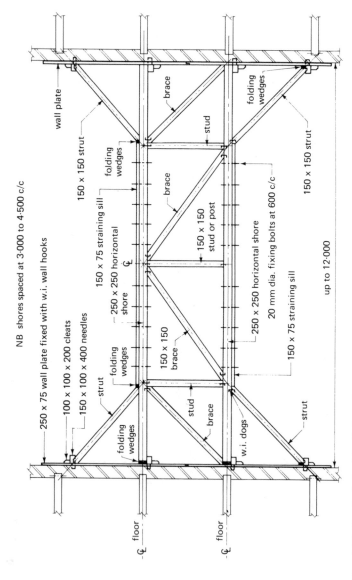

NB shores spaced at 3·000 to 4·500 c/c

250 × 75 wall plate fixed with w.i. wall hooks

wall plate

100 × 100 × 200 cleats

150 × 100 × 400 needles

strut

folding wedges

folding wedges

150 × 75 straining sill

250 × 250 horizontal shore

150 × 150 brace

150 × 150 strut

folding wedges

brace

stud

folding wedges

brace

150 × 150 stud or post

250 × 250 horizontal shore

20 mm dia. fixing bolts at 600 c/c

150 × 150 strut

up to 12·000

150 × 75 straining sill

strut

w.i. dogs

brace

stud

₵ floor

₵ floor

Fig. I.13 Typical double flying shore

It is possible with all forms of shoring to build up the principal members from smaller sections by using bolts and timber connectors, ensuring all butt joints are well staggered to give adequate rigidity. This in effect is a crude form of laminated timber construction.

The site operations for the setting out and erection of a flying shoring system are similar to those enumerated for raking shoring.

4

Scaffolding

A scaffold is a temporary structure from which persons can gain access to a place of work in order to carry out building operations, it includes any working platforms, ladders and guard rails. Basically there are two forms of scaffolding:

1. Putlog scaffolds.
2. Independent scaffolds.

PUTLOG SCAFFOLDS

This form of scaffolding consists of a single row of uprights or standards set away from the wall at a distance which will accommodate the required width of the working platform. The standards are joined together with horizontal members called ledgers and are tied to the building with cross members called putlogs. The scaffold is erected as the building rises and is mostly used for buildings of traditional brick construction (see Fig. 1.14).

INDEPENDENT SCAFFOLDS

An independent scaffold has two rows of standards which are tied by cross members called transoms. This form of scaffold does not rely upon the building for support and is therefore suitable for use in conjunction with framed structures (see Fig. 1.15).

Every scaffold should be securely tied to the building at intervals of approximately 3.600 m vertically and 6.000 m horizontally. This can be achieved by using a horizontal tube called a bridle bearing on the inside of the wall and across a window opening with cross members connected to it (see Fig. I.14); alternatively a tube with a reveal pin in the opening can provide a connection point for the cross members (see Fig. I.15). If suitable openings are not available then the scaffold should be strutted from the ground using raking tubes inclined towards the building.

MATERIALS
Scaffolding can be of:

1. Tubular steel.
2. Tubular aluminium alloy.
3. Timber.

Tubular steel
British Standard 1139 gives recommendations for both welded and seamless steel tubes of 48 mm outside diameter with a nominal 38 mm bore diameter. Steel tubes can be obtained galvanised (to guard against corrosion); ungalvanised tubes will require special care such as painting, varnishing or an oil bath after use. Steel tubes are nearly three times heavier than comparable aluminium alloy tubes but are far stronger and since their deflection is approximately one third of aluminium alloy tubes, longer spans can be used.

Aluminium alloy
Seamless tubes of aluminium alloy with a 48 mm outside diameter are specified in BS 1139 for metal scaffolding. No protective treatment is required unless they are to be used in contact with materials such as damp lime, wet cement and sea water, which can cause corrosion of the aluminium alloy tubes. A suitable protective treatment would be to coat the tubes with bitumastic paint before use.

Timber
The use of timber as a temporary structure in the form of a scaffold is now rarely encountered in this country, although it is still used extensively in other countries. The timber used is fir of structural quality in either putlog or independent format, the members being lashed together with wire or rope instead of the coupling fittings used with metal scaffolds.

Scaffold boards
These are usually boards of softwood timber complying with the recommendations of BS 2482 used to form the working platform at the required level. They should be formed out of specified

softwoods of 225 × 38 section and not exceeding 4.800 m in length. To prevent the ends from splitting they should be end bound with not less than 25 mm wide × 0.9 mm galvanised hoop iron extending at least 150 mm along each edge and fixed with a minimum of two fixings to each end. The strength of the boards should be such that they can support a uniformity distributed load of 6.7 kN/m² when supported at 1.200 m centres.

Scaffold fittings

Fittings of either steel or aluminium alloy are covered by the same British Standard as quoted above for the tubes. They can usually be used in conjunction with either tubular metal unless specified differently by the manufacturer. The major fittings used in metal scaffolding are:

Double coupler: the only real load bearing fitting used in scaffolding and is used to join ledgers to standards.

Swivel coupler: composed of two single couplers riveted together so that it is possible to rotate them and use them for connecting two scaffold tubes at any angle.

Putlog coupler: used solely for fixing putlogs or transoms to the horizontal ledgers.

Base plate: a square plate with a central locating spigot used to distribute the load from the foot of a standard on to a sole plate or firm ground. Base plates can also be obtained with a threaded spigot and nut for use on sloping sites to make up variations in levels.

Split joint pin: a connection fitting used to joint scaffold tubes end to end. A centre bolt expands the two segments which grip on the bore of the tubes.

Reveal pin: fits into the end of a tube to form an adjustable strut.

Putlog end: a flat plate which fits on the end of a scaffold tube to convert it into a putlog.

Typical examples of the above fittings are shown in Fig. I.16.

THE CONSTRUCTION (WORKING PLACES) REGULATIONS 1966

This statutory instrument is designed to ensure that suitable and sufficient safe access to and egress from every

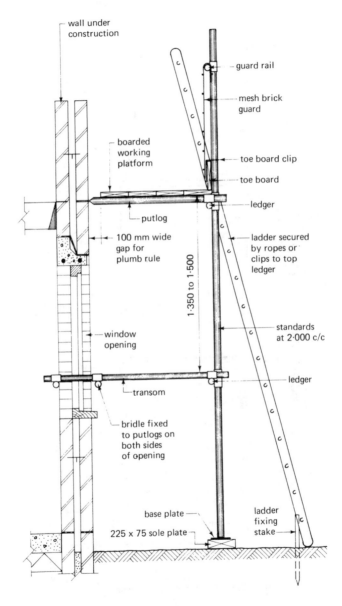

wall under construction

guard rail

mesh brick guard

boarded working platform

toe board clip

toe board

ledger

putlog

100 mm wide gap for plumb rule

ladder secured by ropes or clips to top ledger

1·350 to 1·500

window opening

standards at 2·000 c/c

ledger

transom

bridle fixed to putlogs on both sides of opening

base plate

ladder fixing stake

225 x 75 sole plate

Fig. I.14 Typical tubular steel putlog scaffold

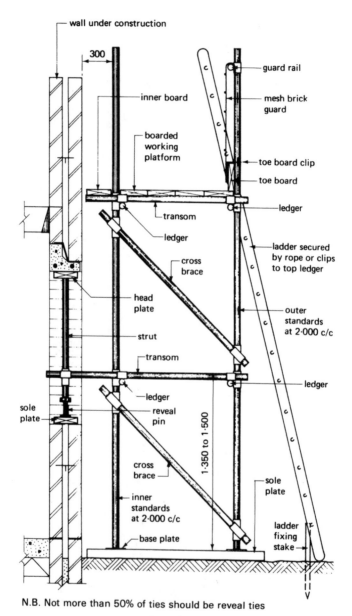

wall under construction

300

inner board

boarded
working
platform

transom

ledger

cross
brace

head
plate

strut

transom

ledger

sole
plate

reveal
pin

1·350 to 1·500

cross
brace

inner
standards
at 2·000 c/c

base plate

guard rail

mesh brick
guard

toe board clip

toe board

ledger

ladder secured
by rope or clips
to top ledger

outer
standards
at 2·000 c/c

ledger

sole
plate

ladder
fixing
stake

N.B. Not more than 50% of ties should be reveal ties

Fig. I.15 Typical tubular steel independent scaffold

30

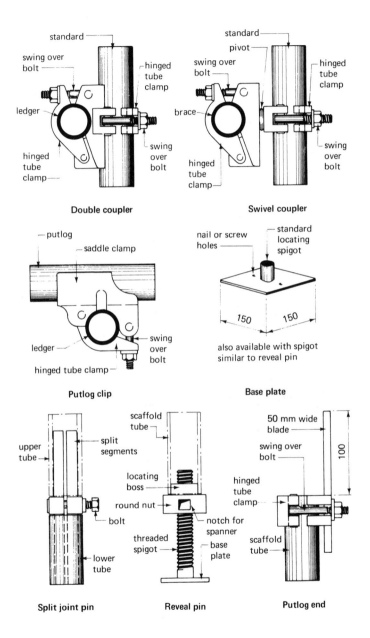

Fig. I.16 Typical steel scaffold fittings

31

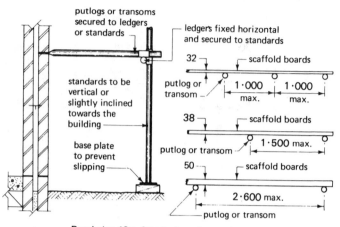

putlogs or transoms secured to ledgers or standards

ledgers fixed horizontal and secured to standards

standards to be vertical or slightly inclined towards the building

base plate to prevent slipping

32 ⌐ scaffold boards
putlog or transom ⌐ 1·000 max. | 1·000 max.

38 ⌐ scaffold boards
putlog or transom ⌐ 1·500 max.

50 ⌐ scaffold boards
2·600 max.
putlog or transom

Regulation 13 ~ Standards, putlogs and transoms

working platform

piers of loose bricks

600 max.

Regulation 15 ~ Stability of scaffolds

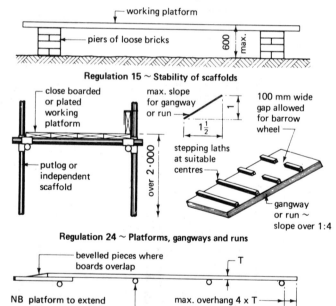

close boarded or plated working platform

putlog or independent scaffold

over 2·000

max. slope for gangway or run

1
1½

stepping laths at suitable centres

100 mm wide gap allowed for barrow wheel

gangway or run ~ slope over 1:4

Regulation 24 ~ Platforms, gangways and runs

bevelled pieces where boards overlap

T

NB platform to extend 600 mm beyond end of working face wherever practicable

boards evenly supported on at least 3 supports per board length

max. overhang 4 x T

Regulation 25 ~ Boards in working platforms

Fig. I.17 Scaffolds and Construction Regulations – 1

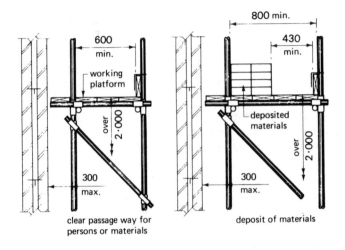

clear passage way for
persons or materials

deposit of materials

Regulation 26 ~ Widths of working platforms for
putlog and independent scaffolds

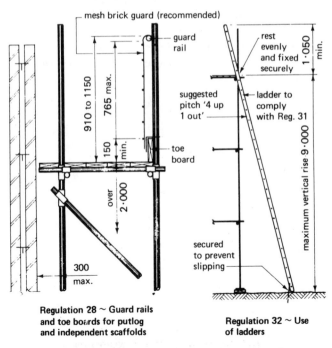

Regulation 28 ~ Guard rails
and toe boards for putlog
and independent scaffolds

Regulation 32 ~ Use
of ladders

Fig. I.18 Scaffolds and Construction Regulations – 2

place at which any person at any time works is provided and properly maintained. Scaffolds and ladders are covered by this document which sets out the minimum requirements for materials, maintenance, inspection and construction of these working places. The main constructional requirements of these regulations are illustrated in Figs. I.17 and I.18. The metric dimensions shown in the figures are those quoted in the Statutory Instrument entitled *The Construction (Metrication) Regulations 1984* as being acceptable metric equivalents of the imperial dimensions given in the actual regulations.

Regulation 22 sets out the requirements for the inspection of scaffolds, by a competent person, which are:

1. Within the preceding seven days.
2. After adverse weather conditions which may have affected the scaffold's strength or stability.

A record of all such inspections must be kept in accordance with Regulation 39 and must give the following information:

1. Location and description of scaffold.
2. Date of inspection.
3. Result of inspection, stating the condition of the scaffold.
4. Signature of person who made the inspection.

The importance of providing a safe and reliable scaffold from which to carry out building operations cannot be over emphasised, since badly constructed and non-maintained scaffolds are a large contributory factor to the high accident rate which prevails in the building industry today.

5
Retaining walls

Part II
Substructures

The basic function of a retaining wall is to retain soil at a slope which is greater than it would naturally assume, usually at a vertical or near vertical position. The natural slope taken up by any soil is called its angle of repose and is measured in relationship to the horizontal. Angles of repose for different soils range from $45°$ to near $0°$ for wet clays but for most soils an average angle of $30°$ is usually taken. It is the wedge of soil resting on this upper plane of the angle of repose which a retaining wall has to support. The walls are designed to offer the necessary resistance by using their own mass to resist the thrust or relying upon the principles of leverage. The terminology used in retaining wall construction is shown in Fig. II.1.

DESIGN PRINCIPLES

The design of any retaining wall is basically concerned with the lateral pressures of the retained soil and any subsoil water. It must be designed to ensure that:

1. Overturning does not occur.
2. Sliding does not occur.
3. The soil on which the wall rests is not overloaded.
4. The materials used in construction are not overstressed.

It is difficult to accurately define the properties of any soil since they are variable materials and the calculation of pressure exerted at any point on the wall is a task for the expert who must take into account the following factors:

1. Nature and type of soil.

2. Height of water table.

3. Subsoil water movements.

4. Type of wall.

5. Materials used in the construction of the wall.

The actual design and calculations for the various types of retaining walls are beyond the scope of this book, but it is essential that a student of construction technology has an appreciation of the factors and considerations involved in retaining wall design.

EARTH PRESSURES

The designer is mainly concerned with the effect of two forms of earth pressure:

1. Active earth pressure.

2. Passive earth pressure.

Active earth pressures are those which at all times are tending to move or overturn the retaining wall and is composed of the earth wedge being retained together with any hydrostatic pressure caused by the presence of ground water. The latter can be reduced by the use of subsoil drainage behind the wall or by inserting drainage openings called weep holes through the thickness of the stem enabling the water to drain away.

Passive earth pressures are reactionary pressures which will react in the form of a resistance to movement of the wall. If the wall tends to move forward the earth in front of the toe will be compressed and a reaction in the form of passive pressure will build up in front of the toe to counteract the forward movement. This pressure can be increased by enlarging the depth of the toe or by forming a rib on the underside of the base. Typical examples of these pressures are shown in Fig. II.1.

STABILITY

The overall stability of a retaining wall is governed by the result of the action and reaction of a number of loads:

Applied loads: such as soil and water pressure on the back of the wall; the mass of the wall and in the case of certain forms of cantilever walls the mass of the soil acting with the mass of the wall.

Induced loads: such as the ground pressure under the base, the passive pressure at the toe and the friction between the underside of the base and the soil.

Effects of water

Ground water behind a retaining wall whether static or percolating through a subsoil can have adverse effects

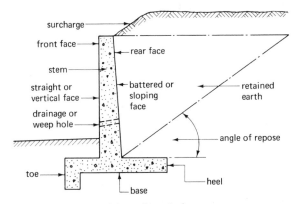

surcharge

front face

rear face

stem

straight or
vertical face

battered or
sloping
face

retained
earth

drainage or
weep hole

angle of repose

toe

heel

base

Retaining wall terminology

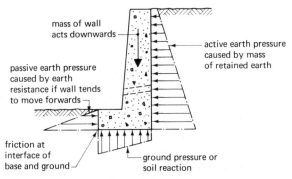

mass of wall
acts downwards

active earth pressure
caused by mass
of retained earth

passive earth pressure
caused by earth
resistance if wall tends
to move forwards

friction at
interface of
base and ground

ground pressure or
soil reaction

Mass retaining walls

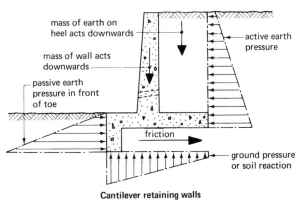

mass of earth on
heel acts downwards

active earth
pressure

mass of wall acts
downwards

passive earth
pressure in front
of toe

friction

ground pressure
or soil reaction

Cantilever retaining walls

Fig. II.1 Retaining wall terminology and pressures

37

upon the design and stability. It will increase the pressure on the back of the wall and by reducing the soil shear strength it can reduce the bearing capacity of the soil; it can reduce the frictional resistance between the base and the soil and reduce the possible passive pressure in front of the wall. It follows therefore that the question of drainage of the water behind the retaining wall is of the utmost importance in the design.

Slip circle failure

This is a form of failure sometimes encountered with retaining walls in clay soils particularly if there is a heavy surcharge. It takes the form of a rotational movement of the soil and wall along a circular arc which starts behind the wall and passes under the base, resulting in a tilting and forward movement of the wall. If the movement is not unacceptable further movement can be prevented by driving sheet piles into the ground in front of the toe to a depth that will cut the slip circle arc.

TYPES OF WALLS

Mass retaining walls

Sometimes called gravity walls and relying upon their own mass together with the friction on the underside of the base to overcome the tendency to slide or overturn. They are generally only economic up to a height of 1.800 m. Mass walls can be constructed of semi-engineering quality bricks bedded in a 1 : 3 cement mortar or of mass concrete. The latter could have some light fabric reinforcement to control surface cracking. Natural stone is suitable for small walls up to 1.000 m high but generally it is used as a facing material for walls over 1.000 m. Typical examples of this are shown in Fig. II.2.

Cantilever walls

Usually of reinforced concrete and work on the principles of leverage. Two basic forms can be considered, a base with a large heel so that the mass of earth above can be added to the mass of the wall for design purposes, or if this form is not practicable a cantilever wall with a large toe must be used (see Fig. II.3). The drawings show typical sections and patterns of reinforcement encountered with these basic forms of cantilever retaining walls. The main steel occurs on the tension face of the wall and nominal steel (0.15% of the cross sectional area of the wall) is very often included in the opposite face to control the shrinkage cracking which occurs in *in situ* concrete work. Reinforcement

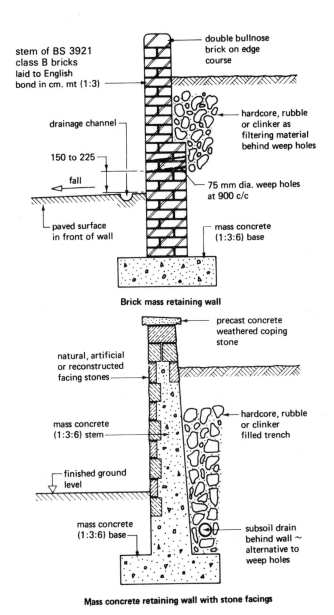

stem of BS 3921
class B bricks
laid to English
bond in cm. mt (1:3)

double bullnose
brick on edge
course

drainage channel

150 to 225

fall

paved surface
in front of wall

hardcore, rubble
or clinker as
filtering material
behind weep holes

75 mm dia. weep holes
at 900 c/c

mass concrete
(1:3:6) base

Brick mass retaining wall

precast concrete
weathered coping
stone

natural, artificial
or reconstructed
facing stones

mass concrete
(1:3:6) stem

finished ground
level

mass concrete
(1:3:6) base

hardcore, rubble
or clinker
filled trench

subsoil drain
behind wall ~
alternative to
weep holes

Mass concrete retaining wall with stone facings

Fig. II.2 Typical mass retaining walls

requirements, bending, fabricating and placing are dealt with in detail in the section on reinforced concrete.

Reinforced cantilever walls have an economic height range of 1.200 to 6.000 m; walls in excess of this height have been economically constructed using prestressing techniques. Any durable facing material may be applied to the surface to improve the appearance of the wall but it must be remembered that such finishes are decorative and add nothing to the structural strength of the wall.

Counterfort retaining walls

These walls can be constructed of reinforced or prestressed concrete and are considered suitable if the height is over 4.500 m. The counterforts are triangular beams placed at suitable centres behind the stem and above the base to enable the stem and base to act as slabs spanning horizontally over or under the counterforts. Fig. II.4 shows a typical section and pattern of reinforcement for a counterfort retaining wall.

If the counterforts are placed on the face of the stem they are termed buttresses and the whole arrangement is called a buttress retaining wall. In both formats the design and construction principles are similar.

Precast concrete retaining walls

Manufactured from high grade precast concrete on the cantilever principle usually to a 600 mm wide module (see Fig. II.5). They can be erected on a foundation as a permanent retaining wall or be free standing to act as a dividing wall between heaped materials such as aggregates for concrete. In the latter situation they can increase by approximately three times the storage volume for any given area. Other advantages are a reduction in time by eliminating the curing period which is required for *in situ* walls and eliminating the need for costly formwork together with the time required to erect and dismantle the temporary forms. The units are reinforced on both faces to meet all forms of stem loading. Lifting holes are provided which can be utilised as strap fixing holes if required. Special units to form internal angles, external angles, junctions and curved walls are also available to provide flexible layout arrangements.

Precast concrete crib retaining walls

Crib walls are designed on the principle of a mass retaining wall. They consist of a framework or crib of precast

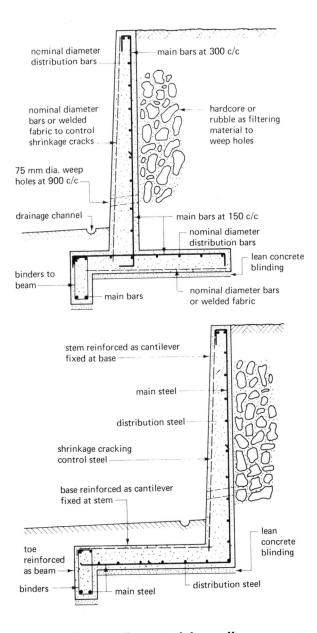

nominal diameter distribution bars

main bars at 300 c/c

nominal diameter bars or welded fabric to control shrinkage cracks

hardcore or rubble as filtering material to weep holes

75 mm dia. weep holes at 900 c/c

drainage channel

main bars at 150 c/c

nominal diameter distribution bars

lean concrete blinding

binders to beam

main bars

nominal diameter bars or welded fabric

stem reinforced as cantilever fixed at base

main steel

distribution steel

shrinkage cracking control steel

base reinforced as cantilever fixed at stem

lean concrete blinding

toe reinforced as beam

binders

main steel

distribution steel

Fig. II.3 Typical R.C. cantilever retaining walls

41

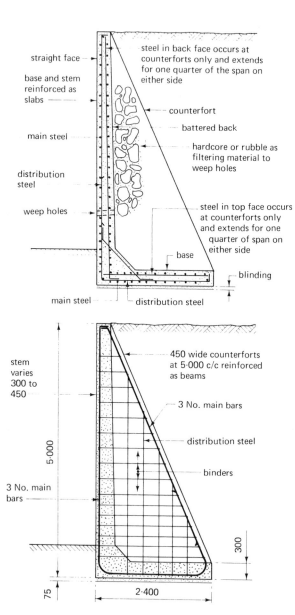

Fig. II.4 Typical R.C. counterfort retaining wall

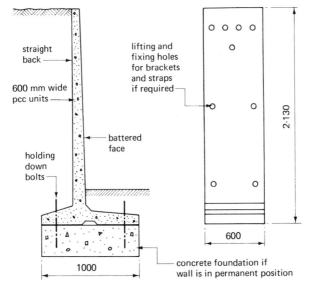

Typical 'Marley' precast concrete retaining wall

straight back

600 mm wide pcc units

battered face

holding down bolts

lifting and fixing holes for brackets and straps if required

2·130

600

1000

concrete foundation if wall is in permanent position

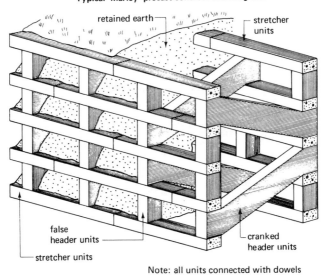

retained earth

stretcher units

false header units

stretcher units

cranked header units

Note: all units connected with dowels

'Anda-Crib' precast concrete retaining wall

Fig. II.5 Precast concrete retaining walls

concrete or timber units within which the soil is retained. They are constructed with a face batter of between 1 : 6 and 1 : 8 unless the height is less than the width of the crib ties when the face can be constructed vertical. Subsoil drainage is not required since the open face provides for adequate drainage (see Fig. II.5).

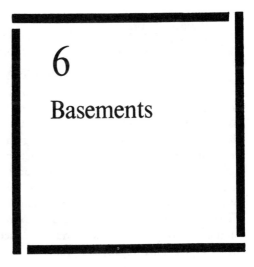

6

Basements

Appendix L in Approved Document B of the Building Regulations 1985 defines a basement storey as a storey with a floor which at some point is more than 1.200 m below the highest level of ground adjacent to the outside walls. This definition is given in the context of inhibiting the spread of fire within a building and generally the fire resistance requirements for basements are more onerous than the ground or upper storeys in the same building. This section on basements is only concerned with basement storeys which are below ground level.

The structural walls of a basement below ground level are in fact retaining walls which have to offer resistance to the soil and ground water pressures as well as assisting to transmit the superstructure loads to the foundations. It is possible to construct a basement free of superstructural loadings but these techniques are beyond the scope of this book.

WATERPROOFING

Apart from the structural design of the basement walls and floor, waterproofing presents the greatest problem in basement construction. Building Regulation C4 requires such walls to be constructed so that they will not transmit moisture from the ground to the inside of the building or to any material used in the construction that would adversely be affected by moisture. Building Regulation C4 also imposes similar conditions on the construction of floors. Basement structures can be waterproofed by one of three basic methods:

1. Monolithic structures.

2. Drained cavities.
3. Membranes.

Monolithic structures

These are basements of dense reinforced concrete using impervious aggregates for the walls and floor to form the barrier to water penetration. Great care must be taken with the design of the mix, the actual mixing and placing together with careful selection and construction of the formwork if a satisfactory water barrier is to be achieved. Shrinkage cracking can largely be controlled by forming construction joints ar regular intervals. These joints should provide continuity of reinforcement and by the incorporation of a PVC or rubber water bar a barrier to the passage of water; typical examples are shown in Fig. II.6. Monolithic structures, whilst providing an adequate barrier to the passage of water, are not always vapourproof.

Drained cavities

This method provides an excellent barrier to moisture penetration of basements by allowing any moisture which has passed through the structural wall to drain down within a cavity formed between the inner face of the structural wall and an inner non-load bearing wall. This internal wall is built of a floor covering of special triangular precast concrete tiles which allows the moisture, from the cavity, to flow away under the tiles to a sump where it is discharged into a drainage system either by gravity or pumping. This method of waterproofing is usually studied in detail during advance courses in construction technology.

Membranes

A membrane is a relatively thin material placed either on the external or internal face of a basement wall or floor to provide the resistance to the passage of moisture to the inside of the basement. If the membrane is applied externally protection is also given to the structural elements and the hydrostatic pressure will keep it firmly in place, but a reasonable working space must be allowed around the perimeter of the basement. This working space will entail extra excavation and subsequent backfilling after the membrane has been applied. If adequate protection is not given to the membrane it can easily be damaged during the backfilling operation. An internally applied membrane gives no protection to the structural elements and there is the danger that the membrane may be forced away from the surfaces, by water pressure, unless it is adequately loaded. These loading coats will reduce the usable volume within the basement (see Figs. II.7 and II.8).

46

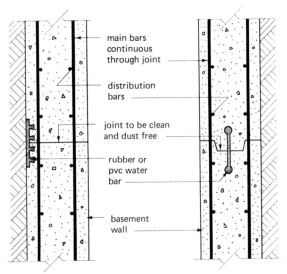

main bars
continuous
through joint

distribution
bars

joint to be clean
and dust free

rubber or
pvc water
bar

basement
wall

Note: horizontal joints positioned at 12 to 15 times
wall thickness

Construction joints

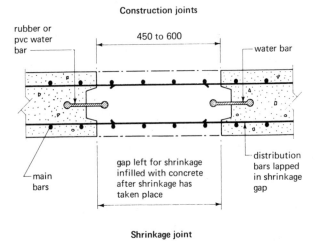

rubber or
pvc water
bar

450 to 600

water bar

gap left for shrinkage
infilled with concrete
after shrinkage has
taken place

distribution
bars lapped
in shrinkage
gap

main
bars

Shrinkage joint

Fig. II.6 Joints and water bars

Suitable materials which can be used for forming membranes are bituminous felt; polythene sheeting; polyisobutylene plastic; epoxy resin compounds; bituminous compounds and mastic asphalt. Asphalt in the form of tanking is the method which is considered in detail in a second year course of study.

Asphalt tanking

Asphalt is a natural or manufactured mixture of bitumen with a substantial proportion of inert mineral matter. When heated, asphalt becomes plastic and can be moulded by hand pressure into any shape. Bitumen is a complex mixture of hydrocarbons and has both waterproofing and adhesive properties. In its natural state asphalt occurs as a limestone rock impregnated with bitumen and is mined notably in France, Switzerland and Sicily. Another source of asphalt is the asphalt lake in Trinidad in the West Indies which was discovered by Sir Walter Raleigh in 1595 and today still yields about 100 000 tonnes annually. In the centre of this lake the asphalt is a liquid but nearer the edges it is a semi-fluid, and although large quantities are removed during the day the lake refills during the night. The lake asphalt is refined or purified in Trinidad and shipped in barrels for use in building construction all over the world. Natural rock asphalt is crushed and processed to remove unwanted mineral matter before being compounded into mastic asphalt. Bitumen for use with mastic asphalt is also made on a large scale as a residue in the distillation of petroleum.

Mastic asphalt is a type of asphalt composed of suitably graded mineral matter and asphaltic cement to form a coherent, voidless and impermeable mass. The asphalt cement consists of bitumen, lake asphalt, asphaltite or blends of these, sometimes with the addition of flux oil, which is used for softening bitumen or rendering it less viscous. The fine aggregates used in mastic asphalt are natural rock asphalt and limestone aggregate which are covered by BS 1162 and 988 respectively. Coarse aggregates of angular stones of igneous and calcareous origin together with naturally occurring graded siliceous material can be added to the compound if required.

The basic principle of asphalt tanking is to provide a continuous waterproof membrane to the base and walls of the basement. Continuity between the vertical and horizontal membranes is of the utmost importance, and since asphalt will set rapidly once removed from the heat source used to melt the blocks, it is applied in layers over small areas; again continuity is the key factor to a successful operation. Joints in successive coats should be staggered by at least 150 mm in horizontal work and at least 75 mm in vertical work.

On horizontal and surfaces up to $30°$ from the horizontal three coats of asphalt should be applied to give a minimum total thickness of 30 mm. Vertical work should also be a three-coat application to give a total thickness of 20 mm. The junction between horizontal and vertical work should be strengthened by a two-coat angle fillet forming a 50 x 50 mm chamfer. To prevent curling and consequent infiltration of moisture behind the vertical tanking the top edge should be turned into a splayed chase or groove 25 mm wide x 25 mm deep.

It is essential that vertical asphalt is suitably keyed to its background. Concrete formed by using sawn boards for the formwork will usually provide an acceptable surface but smooth concrete will need treatment such as bush hammering the surface and washing to remove all loose particles. Alternatively a primer of sand/cement plastic emulsion or pitch/polymer rubber emulsion can be used. Brick walls can be constructed of keyed bricks or the joints can be raked out to a depth of 20 mm as the work proceeds to provide the necessary keyed surface.

During the construction period the asphalt tanking must be protected against damage from impact, following trades and the adverse effects of petrol and oil. Horizontal asphalt tanking coats should be covered with a fine concrete screed at least 50 mm thick as soon as practicable after laying. Vertical asphalt tanking coats should be protected by building a half brick or block wall 30 mm clear of the asphalt; the cavity so formed should be filled with a mortar grout as the work proceeds to ensure perfect interface contact. In the case of internal tanking this protective wall will also act as the loading coat.

Any openings for the passages of pipes or ducts may allow moisture to penetrate unless adequate precautions are taken. The pipe or duct should be primed and coated with three coats of asphalt so that the sleeve formed extends at least 75 mm on either side of the tanking membrane before being placed in the wall or floor. The pipe or duct is connected to the tanking by a two-coat angle fillet (see Fig. II.7).

The main advantages of mastic asphalt as a waterproof membrane are:

1. It is a thermoplastic material and can therefore be heated and reheated if necessary to make it pliable for moulding with a hand float to any desired shape or contour.
2. Durability: bituminous materials have been used in the construction of buildings for over 5 000 years and have remained intact to this day as shown by excavations in Babylonia.
3. Impervious to both water and water vapour.
4. Non-toxic, vermin and rot proof and it is odourless after laying.
5. It is unaffected by sulphates in the soil, which if placed externally will greatly improve the durability of a concrete structure.

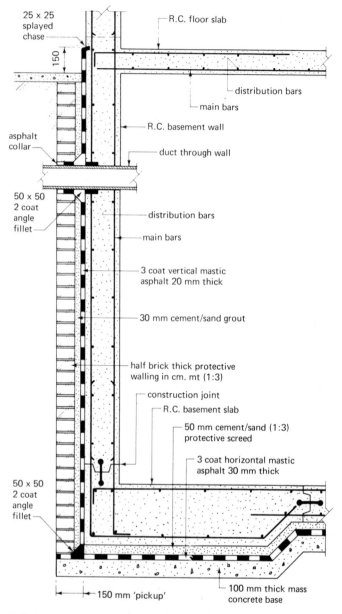

Fig. II.7 External tanking in mastic asphalt

50

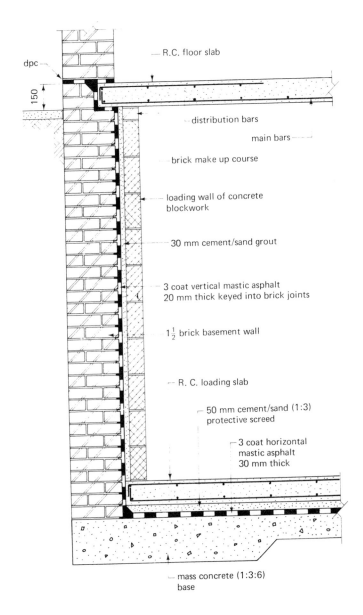

dpc

150

R.C. floor slab

distribution bars

main bars

brick make up course

loading wall of concrete
blockwork

30 mm cement/sand grout

3 coat vertical mastic asphalt
20 mm thick keyed into brick joints

$1\frac{1}{2}$ brick basement wall

R. C. loading slab

50 mm cement/sand (1:3)
protective screed

3 coat horizontal
mastic asphalt
30 mm thick

mass concrete (1:3:6)
base

Fig. II.8 Internal tanking in mastic asphalt

The application of mastic asphalt is recognised as a specialist trade in the building industry and therefore most asphalt work is placed in the hands of specialist sub-contractors, most of which are members of the Mastic Asphalt Council and Employers Federation Limited. The Federation is a non-profit making organisation whose objectives are to provide technical information and promote the use of mastic asphalt as a high quality building material.

7

Reinforced concrete foundations

The function of any foundation is to transmit to the subsoil the loads of the structure. Where a structure has only light loadings such as a domestic dwelling house it is sufficient to use a mass concrete strip foundation or a simple raft. Where buildings are either heavy or transmit the loadings at a series of points such as in a framed building it is uneconomic to use a mass concrete. The plan size of a foundation is a constant feature being derived from:

$$\frac{\text{load}}{\text{bearing capacity of subsoil}}$$

but the thickness of a mass concrete foundation for a heavy load would result in a foundation which is costly and adds unnecessary load to the subsoil. Reinforced concrete foundations are generally cheaper and easier to construct than equivalent mass concrete foundations but will generally require the services of a structural engineer to formulate an economic design. The engineer must define the areas in which tension occurs and specify the reinforcement required, since concrete is a material which is weak in tension. A deeper study of reinforced concrete principles can be found in Part III.

TYPES OF FOUNDATIONS

The principal types of reinforced concrete foundations for buildings are:

1. Strip foundations.

2. Isolated or pad foundations.
3. Raft foundations.
4. Combinations of 1, 2 and 3.
5. Piled foundations.

The foundations listed in 4 and 5 above are generally studied during advance courses in construction technology and are therefore not considered in this volume.

Strip foundations

Reinforced concrete strip foundations are used to support and transmit the loads from heavy walls. The effect of the wall on the relatively thin foundation is to act as a point load and the resultant ground pressure will induce tension on the underside across the width of the strip. Tensile reinforcement is therefore required in the lower face of the strip with distribution bars in the second layer running longitudinally (see Fig. II.9). The reinforcement will also assist the strip in spanning any weak pockets of soil encountered in the excavations.

Isolated or pad foundations

This type of foundation is used to support and transmit the loads from piers and columns. The most economic plan shape is a square but if the columns are close to the site boundary it may be necessary to use a rectangular plan shape of equivalent area. The reaction of the foundation to the load and ground pressures is to cup, similar to a saucer, and therefore main steel is required in both directions. The depth of the base will be governed by the anticipated moments and shear forces, the calculations involved being beyond the scope of this volume. Incorporated in the base will also be the starter bars for a reinforced concrete column or the holding down bolts for a structural steel column (see Fig. II.9).

Raft foundations

The principle of any raft foundation is to spread the load over the entire area of the site. This method is particularly useful where the column loads are heavy and thus requiring large bases or where the bearing capacity is low, again resulting in the need for large bases. Raft foundations can be considered under three headings:

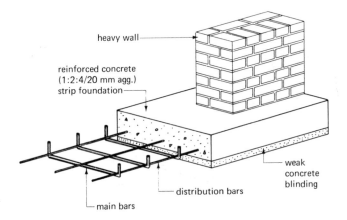

R.C. strip foundation

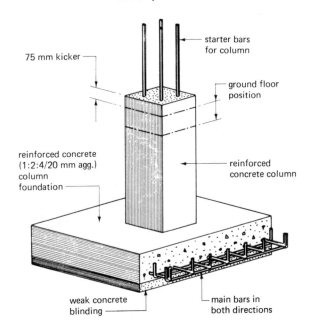

R.C. isolated or pad foundation

Fig. II.9 R.C. strip and pad foundations

55

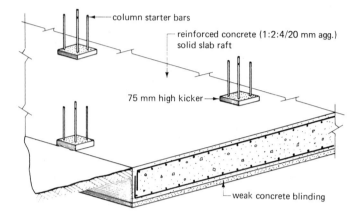

column starter bars

reinforced concrete (1:2:4/20 mm agg.) solid slab raft

75 mm high kicker

weak concrete blinding

R.C. solid slab raft foundation

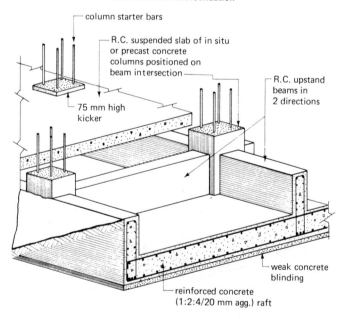

column starter bars

R.C. suspended slab of in situ or precast concrete columns positioned on beam intersection

R.C. upstand beams in 2 directions

75 mm high kicker

weak concrete blinding

reinforced concrete (1:2:4/20 mm agg.) raft

R.C. beam and slab raft foundation

Fig. II.10 R.C. raft foundations

1. Solid slab rafts.
2. Beam and slab rafts.
3. Cellular rafts.

Solid slab rafts are constructed of uniform thickness over the whole raft area, which can be wasteful since the design must be based on the situation existing where the heaviest load occurs. The effect of the load from columns and the ground pressure is to create areas of tension under the columns and areas of tension in the upper part of the raft between the columns. Very often a nominal mesh of reinforcement is provided in the faces where tension does not occur to control shrinkage cracking of the concrete (see Fig. II.10).

Beam and slab rafts are an alternative to the solid slab raft and are used where poor soils are encountered. The beams are used to distribute the column loads over the area of the raft, which usually results in a reduction of the slab thickness. The beams can be upstand or downstand depending upon the bearing capacity of the soil near the surface. Downstand beams will give a saving on excavation costs whereas upstand beams create a usable void below the ground floor if a suspended slab is used (see Fig. II.10).

Cellular rafts

This form of foundation can be used where a reasonable bearing capacity subsoil can only be found at depths where beam and slab techniques become uneconomic. The construction is similar to reinforced concrete basements except that internal walls are used to spread the load over the raft and divide the void into cells. Openings can be formed in the cell walls allowing the voids to be utilised for the housing of services, store rooms or general accommodation (see Fig. II.11).

BLINDING

A blinding layer 50 to 75 mm thick of weak concrete or coarse sand should be placed under all reinforced concrete foundations. The functions of the blinding are to fill in any weak pockets encountered during excavations and to provide a true level surface from which the reinforcement can be positioned. If formwork is required for the foundation some contractors prefer to lay the blinding before

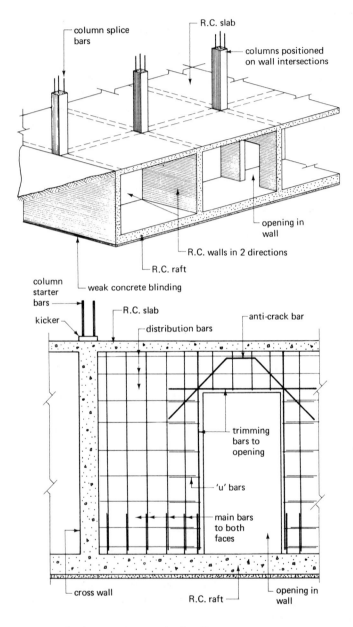

column splice bars

R.C. slab

columns positioned on wall intersections

opening in wall

R.C. walls in 2 directions

R.C. raft

weak concrete blinding

column starter bars

kicker

R.C. slab

distribution bars

anti-crack bar

trimming bars to opening

'u' bars

main bars to both faces

cross wall

R.C. raft

opening in wall

Fig. II.11 Typical cellular raft details

58

assembling the formwork; the alternative is to place the blinding within the formwork and allow this to set before positioning the reinforcement and placing the concrete.

8

Framed buildings

Part III
Simple framed buildings

The purpose of any framed building is to transfer the loads of the structure plus any imposed loads through the members of the frame to a suitable foundation. This form of construction can be clad externally with lightweight non-load bearing walls to provide the necessary protection from the elements and to give the required degree of comfort in terms of sound and thermal insulation. Framed buildings are particularly suitable for medium and high rise structures and for industrialised low rise buildings such as single storey factory buildings.

Frames can be considered under three headings:

Plane frames: fabricated in a flat plane and are usually called trusses or girders according to their elevation shape. They are designed as a series of connected rigid triangles which gives a lightweight structural member using the minimum amount of material; main uses are in roof construction and long span beams of light loading.

Space frames: similar in conception to a plane frame but are designed to span in two directions as opposed to the one-direction spanning of the plane frame. A variation of the space frame is the space deck which consists of a series of linked pyramid frames forming a lightweight roof structure. For details of these forms of framing textbooks on advanced building technology should be consulted.

Skeleton frames: basically these are a series of rectangular frames placed at right angles to one another so that the loads are transmitted from member to member until they are transferred through the foundations to the subsoil. Skeleton frames can be economically constructed of concrete or steel or a combination of the two. Timber skeleton frames, although possible, are generally considered to be uneconomic in this form. The choice of material for a framed structure can be the result of a number of factors such as site conditions, economics, availability of labour and materials, time factor, statutory regulations, capital costs, maintenance costs and personal preference.

Functions of skeleton frame members

Main beams: span between columns and transfer the live and imposed loads placed upon them to the columns.

Secondary beams: span between and transfer their loadings to the main beams. Primary function is to reduce the spans of the floors or roof being supported by the frame.

Tie beams: internal beams spanning between columns at right angles to the direction of the main beams and have the same function as a main beam.

Edge beams: as tie beam but spanning between external columns.

Columns: vertical members which carry the loads transferred by the beams to the foundations.

Foundation: the base(s), to which the columns are connected and serve to transfer the loadings to a suitable load-bearing subsoil.

Floors: may or may not be an integral part of the frame; they provide the platform on which equipment can be placed and on which people can circulate. Besides transmitting these live loads to the supporting beams they may also be required to provide a specific fire resistance, together with a degree of sound and thermal insulation.

Roof: similar to floors but its main function is to provide a weather-resistant covering to the uppermost floor.

Walls: the envelope of the structure which provides the resistance to the elements, entry of daylight, natural ventilation, fire resistance, thermal insulation and sound insulation.

The three major materials used in the construction of skeleton frames, namely, reinforced concrete, precast concrete and structural steel work, are considered in detail in the following chapters.

9

Reinforced
concrete frames

Plain concrete is a mixture of cement, fine aggregate, coarse aggregate and water. Concrete sets to a rock-like mass due to a chemical reaction which takes place between the cement and water, resulting in a paste or matrix which binds the other constituents together. Concrete gradually increases its strength during the curing or hardening period to obtain its working strength in about twenty-eight days if ordinary Portland cement is used. The strength achieved will depend on a number of factors such as:

Type of cement used: in all cases the cement should conform to the appropriate British Standard.

Type and size of aggregates: in general aggregates should comply with the various British Standards and have nominal maximum coarse aggregate sizes of 40, 20, 14 and 10 mm.

Water: this should be clean, free from harmful matter and comply with the requirements of BS 3148.

Use of admixtures: individually admixtures can be used as accelerators, for air entrainment to give a weight reduction and added protection against water penetration, chemical and fungal attack. The instructions given by the manufacturer or engineer must be carefully followed since many admixtures if incorrectly used can have serious adverse effects on the hardened concrete.

Water/cement ratio: a certain amount of water is required to hydrate the cement and any extra water is needed only to produce workability. The workability of fresh concrete should be such that it can be handled, placed and compacted so that it will surround any reinforcement and completely fill the formwork. Workability can be specified in terms of a slump test, compacting factor test or VB consistometer test and should not vary beyond the limits recommended in BS 8110 — the structural use of concrete.

Hardened concrete should be specified by the grade required, which is numerically related to its characteristic strength or cube test strength taken at twenty-eight days for concrete with any type of cement except high alumina cement which hardens rapidly. The grades recommended in BS 8110 and BS 5328 are:

Grades 7 and 10 for plain concrete.
Grade 15 for reinforced concrete with lightweight aggregate.
Grades 20 and 25 for reinforced concrete with dense aggregates.
The characteristic strength of grade 7 concrete is 7.0 N/mm^2, and for grade 10 concrete it is 10.0 N/mm^2; similar values can be deduced for the other grades listed above.

Plain concrete in common with other brittle materials has a greater crushing or compressive strength than tensile strength. The actual ratio varies but plain concrete is generally considered to be ten times stronger in compression than in tension. If a plain concrete member is loaded so that tension is induced it will fail in tension when the compressive strength of concrete has only reached one-tenth of its ultimate value. If this weakness in tension can be reinforced in such a manner that the tensile resistance is raised to a similar value as its compressive strength the member will be able to support a load ten times that of plain concrete or alternatively for any given load a smaller section can be used if concrete is reinforced.

REINFORCEMENT

Any material specified for use as a reinforcement to concrete must fulfil certain requirements if an economic structural member is to be constructed. These basic requirements are:

1. Tensile strength.
2. Must be capable of achieving this tensile strength without undue strain.
3. Be of a material that can be easily bent to any required shape.
4. Its surface must be capable of developing an adequate bond between the concrete and the reinforcement to ensure that the required design tensile strength is obtained.
5. A similar coefficient of thermal expansion is required to prevent

unwanted stresses being developed within the member due to temperature changes.

6. Availability at a reasonable cost which must be acceptable to the overall design concept.

The material which meets all the above requirements is steel in the form of bars, and is supplied in two basic types, namely mild steel and high yield steel. Hot rolled steel bars are covered by BS 4449 which specifies a characteristic strength of 250 N/mm^2 for mild steel and 410 N/mm^2 for high yield steel. The surface of mild steel provides adequate bond but the bond of high yield bars, being more critical with the higher stresses developed, is generally increased by rolling on to the surface of the bar longitudinal or transverse ribs. As an alternative to hot rolled steel bars, cold worked steel bars complying with BS 4461 can be used. When bars are cold worked they become harder, stiffer and develop a higher tensile strength, this being 460 N/mm^2 for bars up to 16 mm nominal diameter and 425 N/mm^2 for bars over 16 mm.

The range of diameters available for both round and deformed bars recommended are 6, 8, 10, 12, 16, 20, 25, 32, and 40 with a recommended maximum length of 12.000 m. For pricing purposes the 16 and 20 mm bars are taken as basic with the diameters on either side becoming more expensive as the size increases or decreases. A good design will limit the range of diameters used together with the type of steel chosen to achieve an economic structure and to ease the site processes of handling, storage, buying and general confusion that can arise when the contractor is faced with a wide variety of similar materials.

The bending of reinforcement can be carried out on site by using a bending machine which shapes the cold bars by pulling them round a mandrel. Small diameters can also be bent round simple jigs such as a board with dowels fixed to give the required profile; large diameters may need a power-assisted bending machine. Bars can also be supplied ready bent and labelled so that only the fabrication processes takes place on site.

The bent reinforcement should be fabricated into cages for columns and beams and into mats for slabs and walls. Where the bars cross or intersect one another they should be tied with soft iron wire, fixed with special wire clips or tack welded to maintain their relative positions. Structural members which require only small areas of reinforcement can be reinforced with steel fabric which can be supplied in sheets or rolls.

Steel fabric for the reinforcement of concrete is covered by BS 4483, which gives four basic preferred types. The fabric is factory-made by welding or interweaving wires complying with the requirements of BS 4482 to form sheets with a length of 4.800 m and a width of 2.400 m or alternatively rolls of 48.000 and 72.000 m in length with a

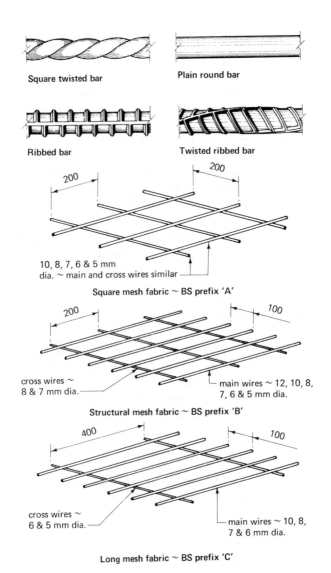

Square twisted bar

Plain round bar

Ribbed bar

Twisted ribbed bar

200   200

10, 8, 7, 6 & 5 mm
dia. ~ main and cross wires similar

Square mesh fabric ~ BS prefix 'A'

200   100

cross wires ~
8 & 7 mm dia.

main wires ~ 12, 10, 8,
7, 6 & 5 mm dia.

Structural mesh fabric ~ BS prefix 'B'

400   100

cross wires ~
6 & 5 mm dia.

main wires ~ 10, 8,
7 & 6 mm dia.

Long mesh fabric ~ BS prefix 'C'

Fig. III.1 Typical reinforcing bars and welded fabric

common width of 2.400 m. Each type has a letter prefix which is followed by a reference number which is the total cross sectional area of main bars in mm^2 per metre width. Typical examples of reinforcing bars and fabric are shown in Fig. III.1.

Before placing reinforcement into the formwork it should be brushed free of all loose rust and mill scale and be free of grease or oil, as the presence of any of these on the surface could reduce the bond and hence the strength of the reinforced concrete. Reinforcement must have a minimum cover of concrete to give the steel protection from corrosion due to contact with moisture and to give the structural member a certain degree of fire resistance. Nominal cover to reinforcement should always be equal to the size of the bar being used or where groups of bars are used at least the size of the largest diameter. BS 8110, Table 3.4 sets out the recommended nominal covers in relationship to the exposure condition and the concrete grade and Approved Document B sets out the minimum fire resistance for various purpose groups of buildings. It must be noted that the dimensions of structural members are also of importance so that failure of the concrete due to the high temperatures encountered during a fire is avoided before the reinforcement reaches its critical temperature. Table A3 of Approved Document B and Table 3.5 and Figure 3.2 in BS 8110 give suitable dimensions for various fire-resistance periods and methods of construction of reinforced concrete members.

To maintain the right amount of concrete cover during construction small blocks of concrete may be placed between the reinforcement and the formwork; alternatively, plastic clips or spacer rings can be used. Where top reinforcement such as in a slab has to be retained in position cradles or chairs made from reinforcing bar may have to be used. All forms of spacers must be of a material which will not lead to corrosion of the reinforcement or cause spalling of the hardened concrete.

Design

The design of reinforced concrete is the prerogative of the structural engineer and is a subject for special study but the technologist should have an understanding of the principles involved. The designer can by assessing the possible dead and live loads on a structural member calculate the reactions and effects such loadings will have on the member. This will enable him to determine where reinforcement is required and how much is needed. He bases his calculations upon the recommendations contained in BS 8110 which is for the structural use of concrete, together with formulae to enable him to determine bending moments, shear forces and the area of steel required. Typical examples of these forces for simple situations are shown in Fig. III.2.

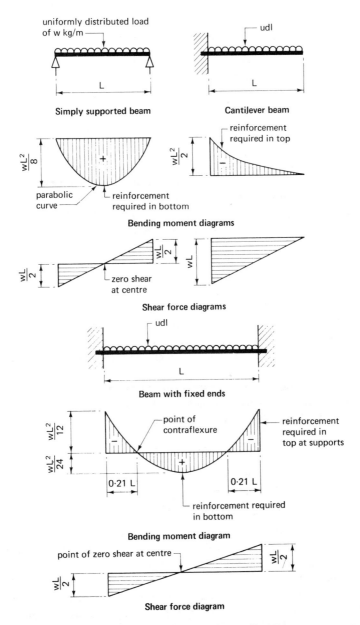

Fig. III.2 **Bending moment and shear force diagrams**

Reinforcement schedules and details

Once the engineer has determined the reinforcement required detail drawings can be prepared to give the contractor the information required to construct the structure. The drawings should give the following information:

1. Sufficient cross reference to identify the member in relationship to the whole structure.
2. All the necessary dimensions for design and fabrication of formwork.
3. Details of the reinforcement.
4. Minimum cover of concrete over reinforcement.
5. Concrete grade required if not already covered in the specification.

Reinforced concrete details should be prepared so that there is a distinct definition between the lines representing the outline of the member and those representing the reinforcement. Bars of a common diameter and shape are normally grouped together with the same reference number when included in the same member. To simplify the reading of reinforced concrete details it is common practice to show only one bar of each group in full together with the last bar position (see Fig. III.3).

The bars are normally bent and scheduled in accordance with the recommendations of BS 4466 which gives details of the common bending shapes, the method of setting out the bending dimensions, the method of calculating the total length of the bar required together with a shape code for use with data processing routines. A preferred form of bar schedule is also given, which has been designed to give easy cross reference to the detail drawing.

Reinforcement on detail drawings is annotated by a coding system to simplify preparation and reading of the details, for example:

9 R 1201 — 300 which can be translated as:

> 9 = total number of bars in the group
> 12 = diameter in mm
> 01 = bar mark number
> 300 = spacing centre to centre
> R = mild steel round bar

The code letter R could be replaced by code letter Y which is used for high yield round bars and high yield square twisted bars; other types of bars are coded with letter X. A typical reinforced concrete beam detail and schedule is shown in Fig. III.3. All other reinforced concrete details shown in this volume are intended to show patterns of reinforcement rather than detailing practice and are therefore shown in full without reference to bar diameters and types.

68

HOOKS, BENDS AND LAPS

To prevent bond failure bars should be extended beyond the section where there is no stress in the bar. The length of bar required will depend upon such factors as grade of concrete, whether the bar is in tension or compression and if the bar is deformed or plain. Hooks and bends can be used to reduce this anchorage length at the ends of bars and should be formed in accordance with the recommendations of BS 4466 (see Fig. III.4).

Where a transfer of stress is required at the end of a bar the bars may be welded or lapped.

BS 8110 recommends that laps and joints should only be made by the methods specified and at the positions shown on the drawings and as agreed by the engineer.

REINFORCED CONCRETE BEAMS

Beams can vary in their complexity of design and reinforcement from the very simple beam formed over an isolated opening such as those shown in Figs. III.3 and III.5 to the more common form encountered in frames where the beams transfer their loadings to the columns (see Fig. III.6).

When tension is induced into a beam the fibres will lengthen until the ultimate tensile strength is reached, when cracking and subsequent failure will occur. With a uniformly distributed load the position and value of tensile stress can easily be calculated by the structural engineer, but the problem becomes more complex when heavy point loads are encountered and this latter situation is considered beyond the scope of a second year course.

The correct design of a reinforced concrete beam will ensure that it has sufficient strength to resist both the compression and tensile forces encountered in the outer fibres, but it can still fail in the 'web' connecting the compression and tension areas. This form of failure is called shear failure and is in fact diagonal tension. Concrete has a limited amount of resistance to shear failure and if this is exceeded reinforcement must be added to provide extra resistance. Shear occurs at or near the supports as a diagonal failure line at an angle of approximately 45° to the horizontal and sloping downwards towards the support. A useful fact to remember is that zero shear occurs at the point of maximum bending (see Fig. III.2).

Reinforcement to resist shearing force may be either stirrups or inclined bars, or both. The total shearing resistance is the sum of the shearing resistances of the inclined bars and the stirrups, calculated separately if both are provided. Inclined or bent up bars should be at 45° to the

69

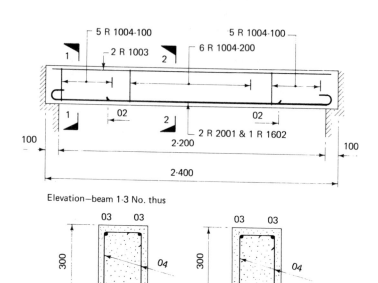

Elevation—beam 1·3 No. thus

'1-1'

'2-2'

NB Cover to main bars 25 mm

Member	Bar mark	Type & size	No. of mbrs	No. in each	Total No.	Length of each bar†	Shape. All dimensions* are in accordance with BS 4466
Beam 1	1	R20	3	2	6	2660	⊂ 2300 ⊃
	2	R16	3	1	3	1400	straight
	3	R10	3	2	6	2300	straight
	4	R10	3	16	48	1000	250 ▢ 150
† specified to nearest 25 mm						* specified to nearest 5 mm	

Fig. III.3 Typical R.C. beam details and schedule

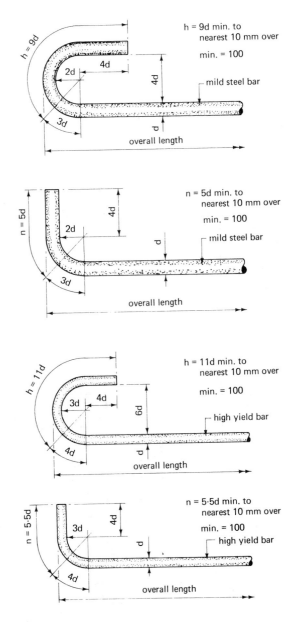

Fig. III.4 Standard hooks and bends

72

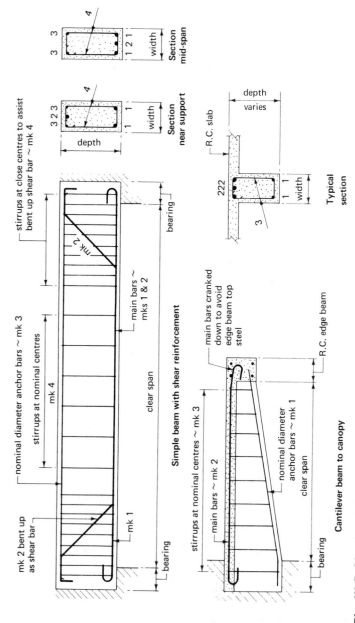

Fig. III.5 Simple reinforced concrete beams

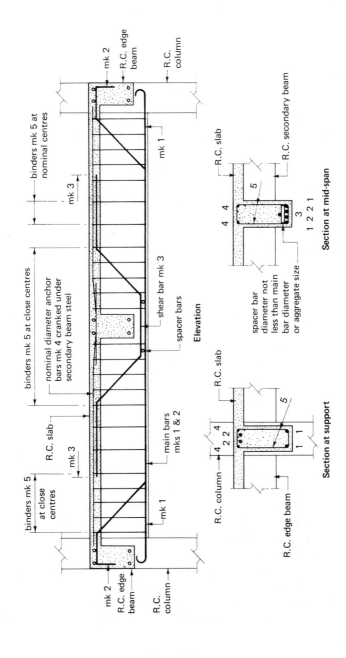

Fig. III.6 R.C. beam with heavy reinforcement

binders mk 5 at nominal centres

mk 2
R.C. edge beam
R.C. column

mk 1

mk 3

binders mk 5 at close centres

nominal diameter anchor bars mk 4 cranked under secondary beam steel

shear bar mk 3

spacer bars

Elevation

R.C. slab

mk 3

main bars mks 1 & 2

mk 1

binders mk 5 at close centres

R.C. slab

mk 2
R.C. edge beam
R.C. column

R.C. slab

R.C. secondary beam

4 4

5

3
1 2 2 1

Section at mid-span

spacer bar diameter not less than main bar diameter or aggregate size

R.C. slab

4 4
2 2

5

1 1

Section at support

R.C. column

R.C. edge beam

73

horizontal and positioned to cut the anticipated shear failure plane at right angles. These may be separate bars or alternatively main bars from the bottom of the beam which are no longer required to resist tension which can be bent up and carried over or onto the support to provide the shear resistance (see Figs. III.5 and III.6). Stirrups or binders are provided in beams, even where not required for shear resistance, to minimise shrinkage cracking and to form a cage for easy handling. The nominal spacing for stirrups must be such that the spacing dimension used is not greater than the lever arm of the section, which is the depth of the beam from the centre of the compression area to the centre of the tension area or 0.75 times the effective depth of the beam, which is measured from the top of the beam to the centre of the tension reinforcement. If stirrups are spaced at a greater distance than the lever arm it would be possible for a shearing plane to occur between consecutive stirrups, but if the centres of the stirrups are reduced locally about the position at which shear is likely to occur several stirrups may cut the shear plane and therefore the total area of steel crossing the shear plane is increased to offer the tensile resistance to the shearing force (see Figs. III.5 and III.6).

REINFORCED CONCRETE COLUMNS

A column is a vertical member carrying the beam and floor loadings to the foundation and is a compression member. Since concrete is strong in compression it may be concluded that provided the compressive strength of the concrete is not exceeded no reinforcement will be required. For this condition to be true the following conditions must exist:

1. Loading must be axial.
2. Column must be short, which can be defined as a column where the ratio of its effective height to its thickness does not exceed 12.
3. Cross section of the column must be large.

These conditions rarely occur in framed buildings, consequently bending is induced and the need for reinforcement to provide tensile strength is apparent. Bending in columns may be induced by one or more of the following conditions:

1. Load coupled with the slenderness of the column; a column is considered to be slender if the ratio of effective height to thickness exceeds 12.
2. Reaction to beams upon the columns, as the beam deflects it tends to pull the column towards itself thus inducing tension in the far face.

3. The reaction of the frame to wind loadings both positive and negative.

The minimum number of main bars in a column should not be less than four for rectangular columns and six for circular columns with a total cross section area of not less than 6% of the cross sectional area of the column and a minimum diameter of 12 mm. To prevent the slender main bars from buckling and hence causing spalling of the concrete, links or binders are used as a restraint. These should be at least one-quarter of the largest main bar diameter and at a pitch or spacing not greater than twelve times the main bar diameter. All bars in compression should be tied by a link passing around the bar in such a way that it tends to move the bar towards the centre of the column; typical arrangements are shown in Fig. III.7.

Where the junction between beams and columns occur there could be a clash of steel since bars from the beam may well be in the same plane as bars in the columns. To avoid this situation one group of bars must be bent or cranked into another plane; it is generally considered that the best practical solution is to crank the column bars to avoid the beam steel; typical examples of this situation together with a method using straight bars are shown in Fig. III.8. A similar situation can occur where beams of similar depth intersect; see cantilever beam example in Fig. III.5.

REINFORCED CONCRETE SLABS

A reinforced concrete slab will behave in exactly the same manner as a reinforced concrete beam and it is therefore designed in the same manner. The designer will analyse the loadings, bending moments, shear forces and reinforcement requirements on a slab strip 1.000 m wide. In practice the reinforcement will be fabricated to form a continuous mat. For light loadings a mat of welded fabric could be used.

There are three basic forms of reinforced concrete slabs, namely:
1. Flat slab floors or roofs.
2. Beam and slab floors or roofs.
3. Ribbed floors or roofs — see Part IV.

Flat slabs

These are basically slabs contained between two plain surfaces and can be either simple or complex. The design of the complex form is based upon the slab acting as a plate in which the slab is divided into middle and column strips; the reinforcement being concentrated in the latter strips. For the purposes of a second year course only the simple flat slab will be dealt with in detail.

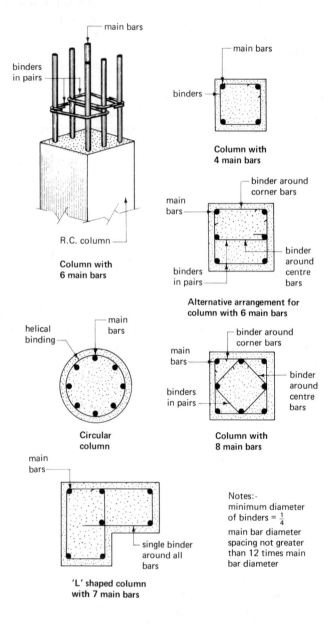

main bars

binders in pairs

R.C. column

Column with
6 main bars

main bars

binders

Column with
4 main bars

binder around
corner bars

main bars

binder around
centre bars

binders in pairs

Alternative arrangement for
column with 6 main bars

helical binding

main bars

Circular
column

binder around
corner bars

main bars

binder around
centre bars

binders in pairs

Column with
8 main bars

main bars

single binder
around all
bars

'L' shaped column
with 7 main bars

Notes:-
minimum diameter
of binders = $\frac{1}{4}$
main bar diameter
spacing not greater
than 12 times main
bar diameter

Fig. III.7 Typical R.C. column binding arrangements

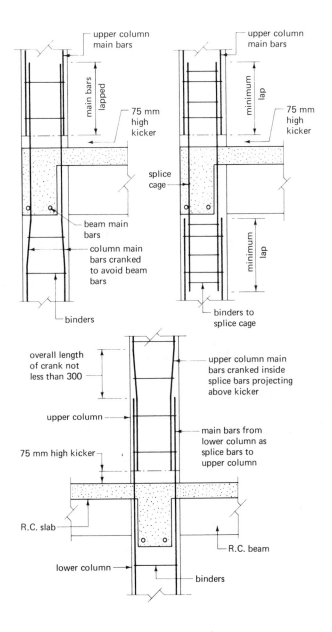

Fig. III.8 R.C. column and beam junctions

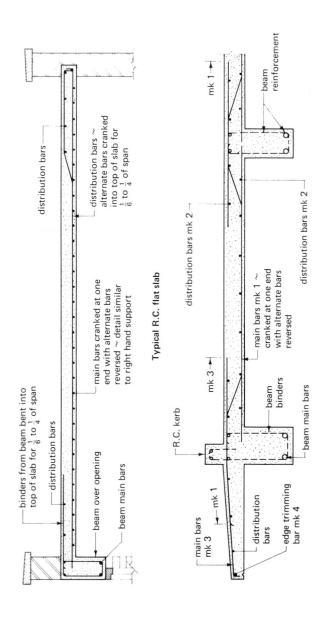

distribution bars

distribution bars ~ alternate bars cranked into top of slab for $\frac{1}{6}$ to $\frac{1}{4}$ of span

main bars cranked at one end with alternate bars reversed ~ detail similar to right hand support

Typical R.C. flat slab

binders from beam bent into top of slab for $\frac{1}{6}$ to $\frac{1}{4}$ of span

distribution bars

beam over opening

beam main bars

mk 1

beam reinforcement

distribution bars mk 2

main bars mk 1 ~ cranked at one end with alternate bars reversed

distribution bars mk 2

mk 3

beam binders

beam main bars

R.C. kerb

main bars mk 3

mk 1

distribution bars

edge trimming bar mk 4

Typical R.C. beam and slab with cantilever

Fig. III.9 Typical R.C. slab details

Simple flat slabs can be thick and heavy but have the advantage of giving clear ceiling heights since there are no internal beams. They are generally economic up to spans of approximately 9.000 m and can be designed to span one way, that is across the shortest span, or to span in two directions. These simple slabs are generally designed to be simply supported, that is, there is no theoretical restraint at the edges and there-fore tension is not induced and reinforcement is not required. However, it is common practice to provide some top reinforcement at the supports as anti-crack steel should there, in practice, be a small degree of restraint. Generally this steel is 50% of the main steel requirement and extends into the slab for 0.2 m of the span. An economic method is to crank up 50% of the main steel or every alternate bar over the support since the bending moment would have reduced to such a degree at this point it is no longer required in the bottom of the slab. If there is an edge beam the top steel can also be provided by extending the beam binders into the slab (see Fig. III.9).

Beam and slab

By adopting this method of design large spans are possible and the reinforcement is generally uncomplicated. A negative moment will occur over the internal supports necessitating top reinforcement; as with the flat slabs, this can be provided by cranked bars (see Fig. III.9). Each bar is in fact cranked but alternate bars are reversed thus simplifying bending and identification of the bars. Alternatively a separate mat of reinforcement supported on chairs can be used over the supports.

10

Formwork

Formworks for *in situ* concrete work may be described as a mould or box into which wet concrete can be poured and compacted so that it will flow and finally set to the inner profile of the box or mould. It is important to remember that the inner profile must be opposite to that required for the finished concrete so if, for example, a chamfer is required on the edge a triangle fillet must be inserted into the formwork.

To be successful in its function formwork must fulfil the following requirements:

1. It should be strong enough to support the load of wet concrete which is generally considered to be approximately 2 400 kg/m^3.

2. It must not be able to deflect under load which would include the loading of wet concrete, self weight and any superimposed loads such as operatives and barrow runs over the formwork.

3. It must be accurately set out; concrete being a fluid when placed, it will take up the shape of the formwork which must therefore be of the correct shape, size and in the right position.

4. It must have grout-tight joints. Grout leakage can cause honeycombing of the surface or produce fins which have to be removed. The making good of defective concrete surfaces is both time consuming and costly. Grout leakage can be prevented by using sheet materials and sealing the joints with flexible foamed polyurethane strip or by using a special self adhesive tape.

5. Form sizes should be designed so that they are the maximum size

which can easily be handled by hand or by a mechanical lifting device.

6. Material must be chosen so that it can be easily fixed using either double-headed nails, round wire nails or wood screws. The common method is to use nails and these should be at least two and a half times the thickness of the timber being nailed, in length.

7. The design of the formwork units should be such that they can easily be assembled and dismantled without any members being trapped.

MATERIALS

The requirements for formwork enumerated above makes timber the most suitable material for general formwork. It can be of board form either wrot or unwrot depending on whether a smooth or rough surface is required.

Softwood boards used to form panels for beam and column sides should be joined together by cross members over their backs at centres not exceeding twenty-four times the board's thickness.

The moisture content of the timber should be between 15 and 20% so that the moisture movement of the formwork is reduced to a minimum. If the timber is dry it will absorb moisture from the wet concrete which could weaken the resultant concrete member. It will also cause the form-work to swell and bulge which could give an unwanted profile to the finished concrete. If timber with a high moisture content is used it will shrink and cup, which could result in open joints and a leakage of grout.

Plywood is extensively used to construct formwork units since it is strong, light and supplied in sheets of 1.200 m wide with standard lengths of 2.400, 2.700 and 3.000 m. The quality selected should be an exterior grade and the thickness related to the anticipated pressures so that the minimum number of strengthening cleats on the back are required.

Chipboard can also be used as a formwork material but because of its lower strength will require more supports and stiffeners. The number of uses which can be obtained from chipboard forms is generally less than plywood, softwood boarding or steel.

Steel forms are generally based upon a manufacturer's patent system and within the constraints of that system are an excellent material. Steel is not so adaptable as timber but if treated with care will give thirty or forty uses, which is approximately double that of similar timber forms.

Mould oils and emulsions

Two defects which can occur on the surface of finished concrete are:

1. **Blow holes:** these are small holes being less than 15 mm in diameter caused by air being trapped between the formwork and the concrete face.
2. **Uneven colour:** this is caused by irregular absorption of water from the wet concrete by the formwork material. A mixture of old and new material very often accentuates this particular defect.

Mould oils can be applied to the inside surface of the formwork to alleviate these defects. To achieve a uniform colour an impervious material or lining is recommended but this will increase the risk of blow holes. Mould oils are designed to overcome this problem when using steel forms or linings by encouraging the trapped air to slide up the face of the formwork. A neat oil will encourage blow holes but will discourage uneven colour, whereas a mould oil incorporating an emulsifying agent will discourage blow holes and reduce uneven colouring. Great care must be taken when applying some mould oils since over oiling may cause some retardation of the setting of the cement. Emulsions are either drops of water in oil or conversely drops of oil in water and are easy to apply but should not be used in conjunction with steel forms since they encourage rusting. It should be noted that generally mould oils and emulsions also act as release agents and therefore it is essential that the oil or emulsion is only applied to the formwork and not to the reinforcement since this may cause a reduction of bond.

Formwork linings

To obtain smooth, patterned or textured surfaces the inside of a form can be lined with various materials such as oil-tempered hardboard, moulded rubber, moulded PVC and glass fibre reinforced polyester; the latter is also available as a complete form mould. When using any form of lining the manufacturer's instructions regarding sealing, fixing and the use of mould oils must be strictly followed to achieve a satisfactory result.

TYPES OF FORMWORK

Foundation formwork

Foundations to a framed building consist generally of a series of isolated bases or pads although if these pads are close together it may be more practicable to merge them together to form a strip. If the subsoil is firm and hard it may be possible to excavate the trench or pit for the foundations to the size and depth required and cast the concrete against the excavated faces. Where this method is not practicable formwork will be required. Side and end panels will be required and these should be firmly strutted against the excavation faces to resist the horizontal pressures of the wet concrete and to retain the formwork in the correct position. Ties will be required across the top of the form as a top restraint and these can be utilised to form the kicker for a reinforced concrete column or as a template for casting in the holding down bolts for precast concrete or structural steel columns (see Fig. III.10).

Column formwork

A column form or box consists of a vertical mould which has to resist considerable horizontal pressures in the early stages of casting. The column box should be located against a 75 mm-high plinth or kicker which has been cast monolithic with the base or floor. The kicker not only accurately positions the formwork but also prevents the loss of grout from the bottom edge of the form. The panels forming the column sides can be strengthened by using horizontal cleats or vertical studs which are sometimes called soldiers. The form can be constructed to the full storey height of the column with cut outs at the top to receive the incoming beam forms. The thickness of the sides does not generally provide sufficient bearing for the beam boxes and therefore the cut outs have a margin piece fixed around the opening to provide extra bearing (see Fig. III.11). It is general practice however to cast the columns up to the underside of the lowest beam soffit and to complete the top of the column at the same time as the beam using make-up pieces to complete the column and receive the beam intersections. The main advantage of casting full height columns is the lateral restraint provided by the beam forms, the disadvantage being the complexity of the formwork involved.

Column forms are held together with collars of timber or metal called yokes in the case of timber and clamps when made of metal. Timber yokes are purpose made whereas steel column clamps are adjustable within the limits of the blades (see Fig. III.12).

The spacing of the yokes and clamps should vary with the anticipated pressures, the greatest pressure occurring at the base of the column box. The actual pressure will vary according to:

1. Rate of placing.
2. Type of mix being used — generally the richer the mix the greater the pressure.
3. Method of placing — if vibrators are used pressures can increase up to 50% over hand placing and compacting.
4. Air temperature — the lower the temperature the slower is the hydration process and consequently higher pressures are encountered.

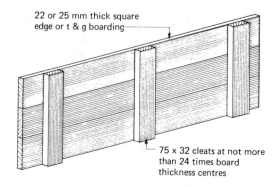

22 or 25 mm thick square edge or t & g boarding

75 x 32 cleats at not more than 24 times board thickness centres

Typical boarded formwork panel

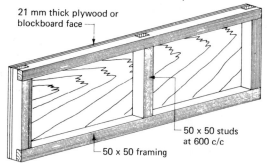

21 mm thick plywood or blockboard face

50 x 50 studs at 600 c/c

50 x 50 framing

Typical framed formwork panel

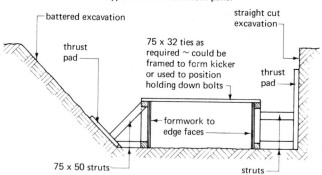

battered excavation

straight cut excavation

thrust pad

75 x 32 ties as required ~ could be framed to form kicker or used to position holding down bolts

thrust pad

formwork to edge faces

75 x 50 struts

struts

Typical foundation formwork

Fig. III.10 Formwork to foundations

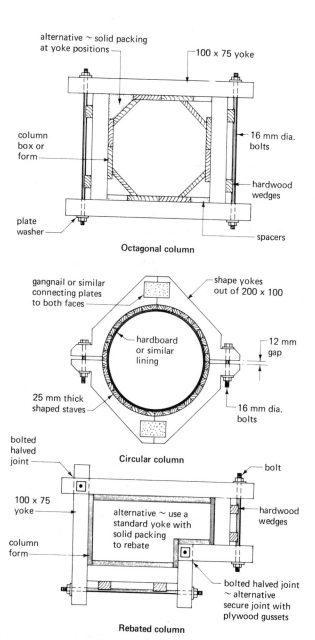

alternative ~ solid packing
at yoke positions

100 x 75 yoke

column
box or
form

16 mm dia.
bolts

hardwood
wedges

plate
washer

spacers

Octagonal column

gangnail or similar
connecting plates
to both faces

shape yokes
out of 200 x 100

hardboard
or similar
lining

12 mm
gap

25 mm thick
shaped staves

16 mm dia.
bolts

Circular column

bolted
halved
joint

bolt

100 x 75
yoke

alternative ~ use a
standard yoke with
solid packing
to rebate

hardwood
wedges

column
form

bolted halved joint
~ alternative
secure joint with
plywood gussets

Rebated column

Fig. III.13 Shaped column forms and yokes

87

Some preliminary raking strutting is required to plumb and align the column forms in all situations. Free standing columns will need permanent strutting until the concrete has hardened but with tied columns the need for permanent strutting must be considered for each individual case.

Shaped columns will need special yoke arrangements unless they are being formed using a patent system. Typical examples of shaped column forms are shown in Fig. III.13.

Beam formwork

A beam form consists of a three-sided box which is supported by cross members called headtrees which are propped to the underside of the soffit board. In the case of framed buildings support to the beam box is also provided by the column form. The soffit board should be thicker than the beam sides since this member will carry the dead load until the beam has gained sufficient strength to be self supporting. Soffit boards should be fixed inside the beam sides so that the latter can be removed at an early date, this will enable a flow of air to pass around the new concrete and speed up the hardening process and also releasing the formwork for reuse at the earliest possible time. Generally the beam form is also used to support the slab formwork and the two structural members are then cast together. The main advantage of this method is that only one concrete operation is involved, although the complexity of the formwork is increased. If the beams and slabs are carried out as separate operations there is the possibility of a shear plane developing between the beam and floor slab; it would be advisable to consult the engineer before adopting this method of construction. Typical examples of beam forms are shown in Fig. III.14.

Structural steelwork has to be protected against corrosion and fire; one method is to encase the steel section with concrete. The steel frame is erected before casting the concrete encasement and in the case of beams it is possible to suspend the form box from the steel section by using a metal device called a hanger fixing or alternatively using a steel column clamp or timber yoke (see Fig. III.15). The hanger fixings are left embedded in the concrete encasing but the bolts and plate washers are recoverable for reuse. If only a haunch is cast around the bottom flange then the projecting hanger fixing wires can be cut off level with the concrete haunch or the floor units can be slotted to receive them.

Slab formwork

Floor or roof slab formwork is sometimes called shuttering and consists of panels of size that can be easily handled.

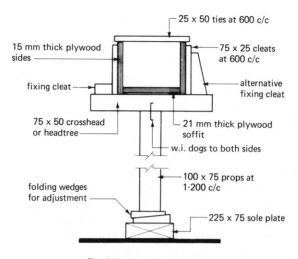

25 x 50 ties at 600 c/c

15 mm thick plywood sides

75 x 25 cleats at 600 c/c

fixing cleat

alternative fixing cleat

75 x 50 crosshead or headtree

21 mm thick plywood soffit

w.i. dogs to both sides

folding wedges for adjustment

100 x 75 props at 1·200 c/c

225 x 75 sole plate

Simple beam or lintel formwork

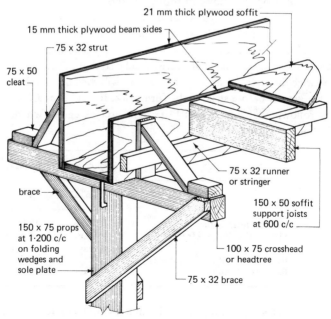

21 mm thick plywood soffit

15 mm thick plywood beam sides

75 x 32 strut

75 x 50 cleat

75 x 32 runner or stringer

brace

150 x 50 soffit support joists at 600 c/c

150 x 75 props at 1·200 c/c on folding wedges and sole plate

100 x 75 crosshead or headtree

75 x 32 brace

Edge beam and slab formwork

Fig. III.14 Typical beam formwork

89

The panels can be framed or joisted and supported by the beam forms with any intermediate propping which is required (see Fig. III.16). Adjustment for levelling purposes can be carried out by using small folding wedges between the joists or framing and the beam box.

SITEWORK

When the formwork has been fabricated and assembled the interior of the forms should be cleared of all rubbish, dirt and grease before the application of any mould oil or releasing agent. All joints and holes should be checked to ensure that they are grout tight.

The distance from the mixer to the formwork should be kept as short as possible to maintain the workability of the mix and to avoid as far as practicable double handling. Care must be taken when placing and compacting the concrete to ensure that the reinforcement is not displaced. The depth of concrete that can be placed in one lift will depend upon the mix and section size. If vibrators are used as the means of compaction this should be continuous during the placing of each batch of concrete until the air expulsion has ceased and care must be taken since over vibrating concrete can cause segregation of the mix.

The striking or removal of formwork should only take place upon instruction from the engineer or agent. The appropriate time at which it is safe to remove formwork can be assessed by tests on cubes taken from a similar batch mixed at the time the concrete was poured and cured under similar conditions. The characteristic cube strength should be 10 N/mm^2 or twice the stress to which the structure will then be submitted whichever is the greater before striking the formwork. If test cubes are not available the following table from BS 8110 can be used as a guide where ordinary Portland cement is used.

Location	Surface or air temperature	
	16° C	7° C
Vertical formwork	12 hours	18 hours
Slab soffits (Props left under)	4 days	6 days
Removal of Props	10 days	15 days
Beam soffits (Props left under)	10 days	15 days
Removal of Props	14 days	21 days

In very cold weather the above minimum periods should be doubled and when using rapid hardening Portland cement the above minimum periods can generally be halved.

Formwork must be removed slowly, as the sudden removal of the

90

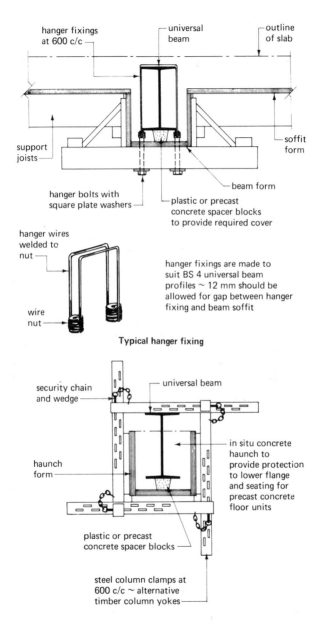

hanger fixings at 600 c/c

universal beam

outline of slab

support joists

soffit form

hanger bolts with square plate washers

beam form

plastic or precast concrete spacer blocks to provide required cover

hanger wires welded to nut

wire nut

hanger fixings are made to suit BS 4 universal beam profiles ~ 12 mm should be allowed for gap between hanger fixing and beam soffit

Typical hanger fixing

security chain and wedge

universal beam

haunch form

in situ concrete haunch to provide protection to lower flange and seating for precast concrete floor units

plastic or precast concrete spacer blocks

steel column clamps at 600 c/c ~ alternative timber column yokes

Fig. III.15 Suspended formwork

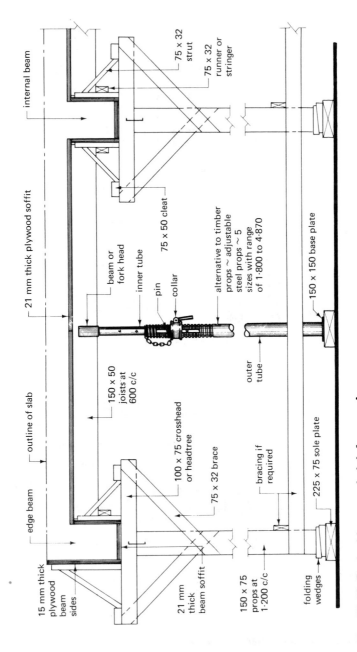

Fig. III.16 Typical beam and slab formwork

internal beam

75 × 32 strut

75 × 32 runner or stringer

21 mm thick plywood soffit

beam or fork head

inner tube

pin

collar

75 × 50 cleat

alternative to timber props ~ adjustable steel props ~ 5 sizes with range of 1·800 to 4·870

150 × 150 base plate

150 × 50 joists at 600 c/c

100 × 75 crosshead or headtree

75 × 32 brace

outer tube

bracing if required

225 × 75 sole plate

outline of slab

edge beam

15 mm thick plywood beam sides

21 mm thick beam soffit

150 × 75 props at 1·200 c/c

folding wedges

wedges is equivalent to a shock load being placed upon the partly hardened concrete. Materials and/or plant should not be placed on the partly hardened concrete without the engineer's permission. When the formwork has been removed it should be carefully cleaned to remove any concrete adhering to the face before being reused. If the forms are not required for immediate reuse they should be carefully stored and stacked to avoid twisting.

The method of curing the concrete will depend upon climatic conditions, type of cement used and the average temperature during the curing period. The objective is to allow the concrete to cure and obtain its strength without undue distortion or cracking. It may be necessary to insulate the concrete by covering with polythene sheeting or an absorbent material which is kept damp to control the surface temperature and prevent the evaporation of water from the surface. Under normal conditions using ordinary Portland cement and with an average air temperature of over 10° C this period would be two days rising to four days during hot weather and days with prolonged drying winds.

11

Precast concrete frames

The overall concept of a precast concrete frame is the same as any other framing material. Single or multi-storey frames can be produced on the skeleton or box frame principle. Single and two-storey buildings can also be produced as portal frames, a method generally reserved for advanced level study. Most precast concrete frames are produced as part of a 'system' building and therefore it is only possible to generalise in an overall study of this method of framing.

Advantages

1. Mixing, placing and curing of the concrete carried out under factory-controlled conditions which results in uniform and accurate units. The casting, being an 'off site' activity, will release site space which would have been needed for the storage of cement and aggregates, mixing position, timber store and fabrication area for formwork and the storage, bending and fabrication of the reinforcement.
2. Repetitive standard units reduce costs: it must be appreciated that the moulds used in precast concrete factories are precision made, resulting in high capital costs. These costs must be apportioned over the number of units to be cast.
3. Frames can be assembled on site in cold weather which helps with the planning, programming and progressing of the building operations. This is important to the contractor since delays can result in the monetary penalty clauses, for late completion of the contract, being invoked.

4. In general the frames can be assembled by semi-skilled labour. With the high turnover rate of labour within the building industry operatives can be recruited and quickly trained to carry out these activities.

Disadvantages

1. System building is less flexible in its design concept than purpose-made structures. It must be noted that there is a wide variety of choice of systems available to the designer, so that most design briefs can be fulfilled without too much modification to the original concept.
2. Mechanical lifting plant will be needed to position the units; this can add to the overall contracting costs since generally larger plant is required for precast concrete structures than for *in situ* concrete structures.
3. Programming may be restricted by controls on delivery and unloading times laid down by the police. Restrictions on deliveries is a point which must be established at the tender period so that the tender programme can be formulated with a degree of accuracy and any overtime payments can be included in the unit rates for pricing.
4. Structural connections between the precast concrete units can present both design and contractual problems. The major points to be considered are protection against weather, fire and corrosion, appearance and the method of construction. The latter should be issued as an instruction to site, setting out in detail the sequence, temporary supports required and full details of the joint.

METHODS OF CONNECTIONS

Foundation connections

Precast columns are connected to their foundations by one of two methods, depending mainly upon the magnitude of the load. For light and medium loads the foot of the column can be placed in a pocket left in the foundation. The column can be plumbed and positioned by fixing a collar around its perimeter and temporarily supporting the column from this collar by using raking adjustable props. Wedges can be used to give added rigidity whilst the column is being grouted into the pocket (see Fig. III.17). The alternative method is to cast or weld on a base plate to the foot of the column and use holding down bolts to secure the column to its foundation in the same manner as described in detail for structural steelwork (see Fig. III.17).

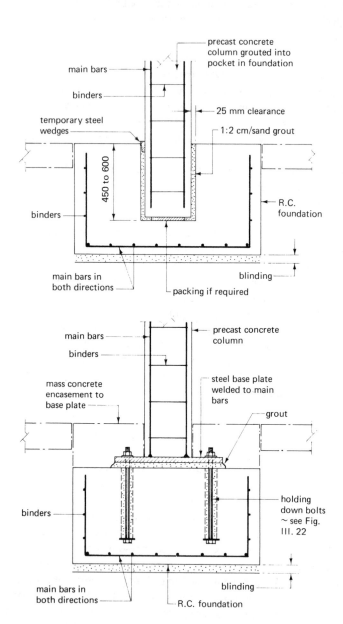

Fig. III.17 P.C.C. column to foundation connections

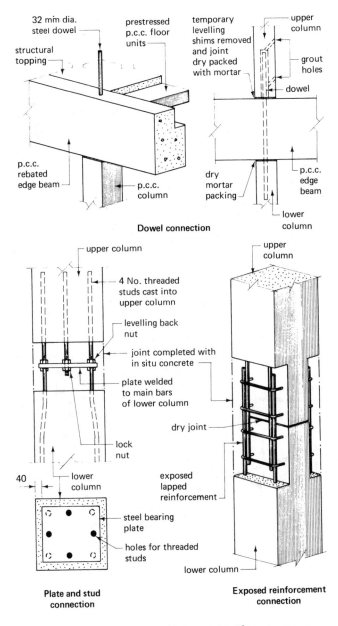

Fig. III.18 Precast concrete column connections

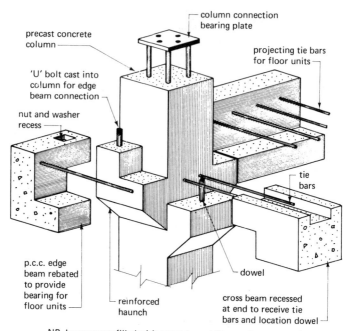

- column connection bearing plate
- precast concrete column
- projecting tie bars for floor units
- 'U' bolt cast into column for edge beam connection
- nut and washer recess
- tie bars
- p.c.c. edge beam rebated to provide bearing for floor units
- dowel
- reinforced haunch
- cross beam recessed at end to receive tie bars and location dowel

NB beam recess filled with cement grout to complete connection

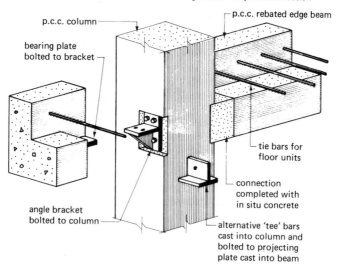

- p.c.c. column
- p.c.c. rebated edge beam
- bearing plate bolted to bracket
- tie bars for floor units
- connection completed with in situ concrete
- angle bracket bolted to column
- alternative 'tee' bars cast into column and bolted to projecting plate cast into beam

Fig. III.19 Typical precast concrete beam connections

98

Column connections

The main principle involved in making column connections is to ensure continuity and this can be achieved by a variety of methods. In simple connections a direct bearing and grouted dowel joint can be used, the dowel being positioned in the upper or lower column. Where continuity of reinforcement is required the reinforcement from both upper and lower columns left exposed and either lapped or welded together before completing the connection with *in situ* concrete. A more complex method is to use a stud and plate connection where one set of threaded bars are connected through a steel plate welded to a set of bars projecting from the lower column; again the connection is completed with *in situ* concrete. Typical column connections are shown in Fig. III.18. Column connections should be made at floor levels but above the beam connections, a common dimension being 600 mm above structural floor level. The columns can be of single or multi-storey height, the latter having provisions for beam connections at the intermediate floor levels.

Beam connections

As with columns, the main emphasis is on continuity within the joint. Three basic methods are used:

1. A projecting concrete haunch is cast on to the column with a locating dowel or stud bolt to fix the beam.
2. A projecting metal corbel is fixed to the column and the beam is bolted to the corbel.
3. Column and beam reinforcement, generally in the form of hooks, are left exposed. The two members are hooked together and covered with *in situ* concrete to complete the joint.

With most beam to column connections lateral restraint is provided by leaving projecting reinforcement from the beam sides to bond with the floor slab or precast concrete floor units (see Fig. III.19).

12

Structural steelwork frames

Structural steel as a means of constructing a framed building has been used since the beginning of the twentieth century and was the major structural material used until the advent of the Second World War, which led to a shortage of the raw material. This shortage led to an increase in the use of *in situ* and precast concrete frames. Today both systems are used and this means a comparison must be made before any particular framing medium is chosen. The main factors to be considered in making this choice are:

Site costs: a building owner will want to obtain a financial return on his capital investment as soon as possible, therefore speed of construction is of paramount importance. The use of a steel or precast concrete frame will enable the maximum amount of prefabrication off site, during which time the general contractor can be constructing the foundations in preparation for the erection of the frame. To obtain the maximum utilisation of a site the structure needs to be designed so that the maximum amount of rentable floor area is achieved. Generally prefabricated section sizes are smaller than comparable *in situ* concrete members, due mainly to the greater control over manufacture obtainable under factory conditions and thus these will occupy less floor area.

Construction costs: the main factors are design considerations, availability of labour, availability of materials and site conditions. Concrete is a flexible material which allows the designer to be more creative than working within the rigid confines of standard steel sections. However, as the complexity of shape and size increases so does the cost of formwork and for the erection of a steel structure skilled labour is required, whereas activities involved with precast concrete structures can be carried out by the more readily available semi-skilled labour working under the direction of a competent person. The availability of materials fluctuates and only a study of current market trends can give an accurate answer to this problem. Site conditions regarding storage space, fabrication areas and manoeuvrability around and over the site can well influence the framing method chosen.

Maintenance costs: these can be considered in the short or long term but it is fair to say that in most framed buildings the costs are generally negligible if the design and workmanship is sound. Steelwork, because of its corrosive properties, will need some form of protective treatment but since most steel structures have to be given a degree of fire resistance the fire protective method may well perform the dual function.

STRUCTURAL STEEL FRAMES

The design, fabrication, supply and erection of a structural steel frame is normally placed in the hands of a specialist sub-contractor. The main contractor's task is to provide the foundation bases in the correct positions and to the correct levels with the necessary holding down fixing bolts. The designer will calculate the loadings, stresses and reactions in the same manner as for reinforced concrete and then select a standard steel member whose section properties meet the design requirements.

Standard steel sections are given in BS 4 and BS 4848 and in the *Handbook on Structural Steelwork* published jointly by the British Constructional Steelwork Association Limited and the Constructional Steel Research and Development Organisation, which gives the following section types:

Universal beams: these are a range of sections supplied with tapered or parallel flanges and are designated by their serial size × mass in kilograms per metre run. To facilitate the rolling operation of universal beam sections the inner profile is a constant dimension for any given serial size. The serial size is therefore only an approximate width and breadth and is given in millimetres.

Joists: a range of small size beams which have tapered flanges and are useful for lintels and small frames around openings. In the case of joists the serial size is the overall nominal dimension.

Universal columns: these members are rolled with parallel flanges and are designated in the same manner as universal beams. It is possible to design a column section to act as a beam and conversely a beam section to act as a column.

Channels: rolled with tapered flanges and designated by their nominal overall dimension x mass per metre run and can be used for trimming and bracing members or as a substitute for joist sections.

Angles: light framing and bracing sections with parallel flanges. The flange or leg lengths can be equal or unequal and the sections are designated by the nominal overall leg lengths x nominal thickness of the flange.

T bars: used for the same purposes as angles and are available as rolled sections with a short or long stalk or alternatively they can be cut from a standard universal beam or column section. Designation is given by the nominal overall breadth and depth x mass per metre run.

Typical standard steel sections are shown in Fig. III.20.

CASTELLATED UNIVERSAL SECTIONS

These are formed by flame cutting a standard universal beam or column section along a castellated line; the two halves so produced are welded together to form an open web beam. The resultant section is one and a half times the depth of the section from which it was cut (see Fig. III.21). This increase in depth gives greater resistance to deflection without adding extra weight but will reduce the clear headroom under the beams unless the overall height of the building is increased. Castellated sections are economical when used to support lightly loaded floor or roof slabs and the voids in the web can be used for housing services. With this form of beam the shear stresses at the supports can be greater than the resistance provided by the web; in these cases one or two voids are filled in by welding into the voids metal blanks.

Connections

Connections in structural steelwork are classified as either shop connections or site connections and can be made by using bolts, rivets or by welding.

Bolts

Black bolts: the cheapest form of bolt available, the black bolt can be either hot or cold forged, the thread being machined onto the shank. The allowable shear stresses for this type of bolt are low and therefore they should only be used for end connections of secondary beams or in

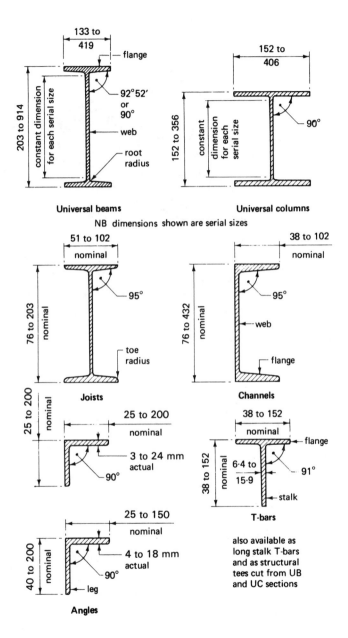

Fig. III.20 Typical BS 4 and BS 4848 steel sections

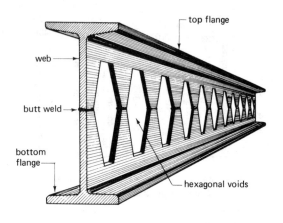

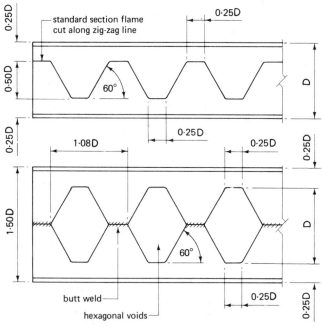

NB castellated joists, universal columns and
zed sections also available

Fig. III.21 Castellated beams

104

conjunction with a seating cleat which has been designed and fixed to resist all the shear forces involved. The clearance in the hole for this form of bolt is usually specified as 1.6 mm over the diameter of the bolt. The term black bolts does not necessarily indicate the colour but is the term used to indicate the comparatively wide tolerances to which these products are usually made. BS 4190 gives recommendations for black bolts and nuts for a diameter range of 5 to 68 mm inclusive.

Bright bolts: these have a machined shank and are therefore of greater dimensional accuracy fitting into a hole with a small clearance allowance. The stresses allowed are similar to those permitted for rivets. Bright bolts are sometimes called turned and fitted bolts.

High strength friction bolts: manufactured from high tensile steels and are used in conjunction with high tensile steel nuts and tempered washers. These bolts have generally replaced rivets and black bolts for both shop and site connections since fewer bolts are needed and hence the connection size is reduced. The object of this form of bolt is to tighten it to a predetermined shank tension in order that the clamping force thus provided will transfer the loads in the connecting members by friction between the parts and not by shear in or bearing on the bolts. Generally a torque controlled spanner or pneumatic impact wrench is used for tightening; other variations to ensure the correct torque are visual indicators such as a series of pips under the head or washer which are flattened when the correct amount of shank tension has been reached. Nominal standard diameters available are from 12 to 36 mm with lengths ranging from 40 to 500 mm, as recommended in BS 4395.

The holes to receive bolts should always be drilled in a position known as the back mark of the section. The back mark is the position on the flange where the removal of material to form a bolt or rivet hole will have the least effect upon the section properties. Actual dimensions and recommended bolt diameters are given in the *Handbook on Structural Steelwork*.

Rivets

Made from mild steel to the recommendations of BS 4620 rivets have been generally superseded by bolted and welded connections for structural steel frames. Rivets are available as either cold or hot forged with a variety of head shapes ranging from an almost semi-circular or snap head to a countersunk head for use when the projection of a snap, universal or flat head would create an obstruction. Small diameter rivets can be cold driven but the usual practice is to drive rivets whilst they are hot. Rivets, like bolts, should be positioned on the back mark of the section; typical spacings are 2½ diameters centre to centre and 1¾ diameters from the end or edge to the centre line of the first rivet.

Welding

Primarily considered as a shop connection since the cost together with the need for inspection, which can be difficult on site, generally makes this method uneconomic for site connections.

The basic methods of welding are oxy-acetylene and electric arc. A blowpipe is used for oxy-acetylene which allows the heat from the burning gas mixture to raise the temperature of the surfaces to be joined. A metal filler rod is held in the flame and the molten metal from the filler rod fuses the surfaces together.

In the alternative method an electric arc is struck between a metal rod connected to a suitable low voltage electrical supply and the surface to be joined which must be earthed or resting on an earthed surface. The heat of the arc causes the electrode or metal rod to melt and the molten metal can be deposited in layers to fuse the pieces to be joined together. With electrical arc welding the temperature rise is confined to the local area being welded whereas oxy-acetylene causes a rise in metal temperature over a general area.

Welds are classified as either fillet or butt welds. Fillet welds are used to the edges and ends of members and forms a triangular fillet of welding material. Butt welds are used on chamfered end to end connections.

Structural steel connections

Base connections: are of one or two forms, the slab or bloom base and the gusset base. In both methods a steel base plate is required to spread the load of the column on to the foundation. The end of the column and the upper surface of the base plate should be machined to give a good inter-face contact when using a bloom base. The base plate and column can be connected together by using cleats or by fillet welding (see Fig. III.22).

The gusset base is composed of a number of members which reduce the thickness of the base plate and can be used to transmit a high bending moment to the foundations. A machined interface between column and base plate will enable all the components to work in conjunction with one another, but if this method is not adopted the connections must transmit all the load to the base plate (see Fig. III.22). The base is joined to the foundation by holding down bolts which must be designed to resist the uplift and tendency of the column to overturn. The bolt diameter, bolt length and size of plate washer are therefore important. To allow for fixing tolerances the bolts are initially housed in a void or pocket which is filled with grout at the same time as the base is grouted on to the foundation. To level and plumb the columns steel wedges are inserted between the underside of the base plate and the top of the foundation (see Fig. III.22).

Beam to column connections: these can be designed as simple connections where the whole of the load is transmitted to the column through a seating cleat. This is an expensive method requiring heavy sections to overcome deflection problems. The usual method employed is the semi-rigid connection where the load is transmitted from the beam to the column by means of top cleats and/or web cleats; for ease of assembly an erection cleat on the underside is also included in the connection detail (see Fig. III.23). A fully rigid connection detail, which gives the greatest economy on section sizes, is made by welding the beam to the column (see Fig. III.23). The uppermost beam connection to the column can be made by the methods described above or alternatively a bearing connection can be used, which consists of a cap plate fixed to the top of the column to which the beams can be fixed either continuously over the cap plate or with a butt joint (see Fig. III.23).

Column splices: these are made at floor levels but above the beam connections. The method used will depend upon the relative column sections (see Fig. III.24).

Beam to beam connections: the method used will depend upon the relative depths of the beams concerned. Deep beams receiving small secondary beams can have a shelf angle connection whereas other depths will need to be connected by web cleats (see Fig. III.25).

FRAME ERECTION

This operation will not normally be commenced until all the bases have been cast and checked since the structural steelwork contractor will need a clear site for manoeuvring the steel members into position. The usual procedure is to erect two storeys of steelwork before final plumbing and levelling takes place.

The grouting of the base plates and holding down bolts is usually left until the whole structure has been finally levelled and plumbed. The grout is a neat cement or cement/sand mixture depending on the gap to be filled:

12 to 25 mm gap — stiff mix of neat cement;

25 to 50 mm gap — fluid mix of 1:2 cement/sand and tamped;

Over 50 mm gap — stiff mix of 1:2 cement/sand and rammed.

With large base plates a grouting hole is sometimes included but with smaller plates three sides of the base plate are sealed with puddle clay,

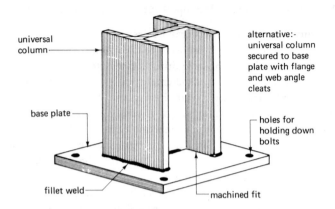

universal column

alternative:- universal column secured to base plate with flange and web angle cleats

base plate

holes for holding down bolts

fillet weld

machined fit

Slab or bloom base

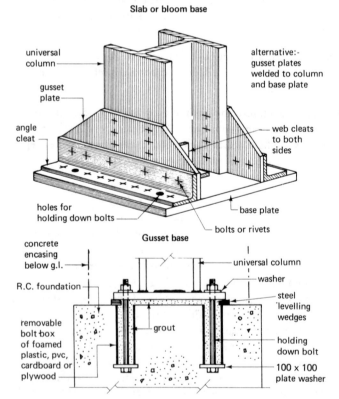

universal column

alternative:- gusset plates welded to column and base plate

gusset plate

angle cleat

web cleats to both sides

holes for holding down bolts

base plate

bolts or rivets

Gusset base

concrete encasing below g.l.

universal column

R.C. foundation

washer

steel levelling wedges

removable bolt box of foamed plastic, pvc, cardboard or plywood

grout

holding down bolt

100 x 100 plate washer

Fig. III.22 Structural steel column bases

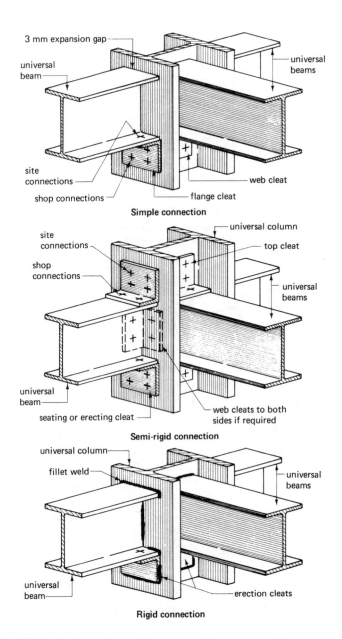

Fig. III.23 Structural steel beam to column connections

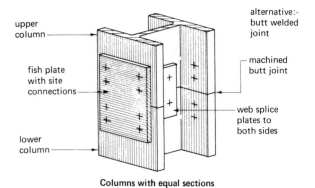

upper column

alternative:-
butt welded joint

fish plate with site connections

machined butt joint

web splice plates to both sides

lower column

Columns with equal sections

NB for columns of same serial size but of different sections splice is made using 4 No. fish plates fixed on the inside of flanges

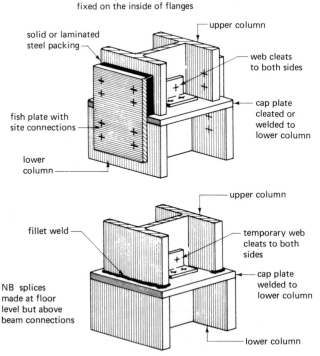

solid or laminated steel packing

upper column

web cleats to both sides

fish plate with site connections

cap plate cleated or welded to lower column

lower column

upper column

fillet weld

temporary web cleats to both sides

cap plate welded to lower column

NB splices made at floor level but above beam connections

lower column

Alternative methods for columns of unequal sections

Fig. III.24 Structural steel column splices

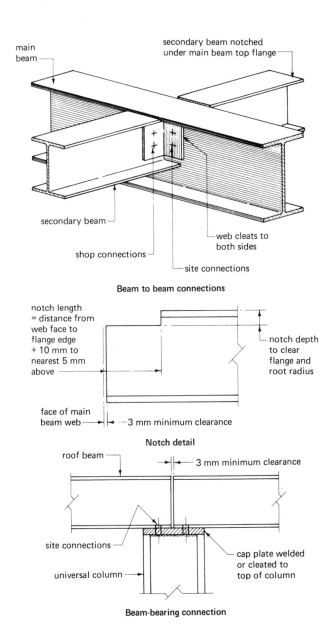

main beam

secondary beam notched under main beam top flange

secondary beam

shop connections

web cleats to both sides

site connections

Beam to beam connections

notch length = distance from web face to flange edge + 10 mm to nearest 5 mm above

notch depth to clear flange and root radius

face of main beam web

3 mm minimum clearance

Notch detail

roof beam

3 mm minimum clearance

site connections

cap plate welded or cleated to top of column

universal column

Beam-bearing connection

Fig. III.25 Structural steel beam to beam connections

bricks or formwork and the grout introduced through the open edge on the fourth side. To protect the base from corrosion it should be encased with concrete up to the underside of the floor level giving a minimum concrete cover of 75 mm to all the steel components.

FIRE PROTECTION OF STEELWORK

Part B of Building Regulations together with Approved Document B gives the minimum fire resistance periods and methods of protection for steel structures according to the purpose group of the building and the function of the member. The traditional method is to encase the steel section with concrete, which requires formwork and adds to the loading of the structure. Many 'dry' techniques are available but not all are suitable for exposed conditions.

13
Claddings

Claddings to buildings can be considered under two classifications:

1. Claddings fixed to a structural backing.
2. Claddings to framed structures.

Claddings fixed to a structural backing

Materials used in this form of cladding are generally considered to be small unit claddings and are applied for one of two reasons. If the structural wall is unable to provide an adequate barrier to the elements a covering of small unit claddings will generally raise the wall's resistance to an acceptable level. Alternatively small unit claddings can be used solely as a decorative feature, possibly to break up the monotony of a large plain area composed of a single material.

The materials used are tiles, slates, shingles, timber boarding, plastic boards and stone facings. The general method of fixing these small units is to secure them to timber battens fixed to the structure backing. Stone and similar facings, however, are usually secured by special mechanical fixings as described later when considering claddings to framed structures.

TILE HANGING
The tiles used in tile hanging can be ordinary plain roofing tiles or alternatively a tile of the same dimensions but having a patterned bottom edge solely for a decorative appearance. The tiles are hung and fixed to tiling battens although nibless tiles fixed directly to the backing wall are

sometimes used (see Fig. III.26). The battens should be impregnated to prevent fungi and insect attack so that their anticipated life is comparable to that of the tiles. Each tile should be twice nailed to its support batten with corrosion resistant nails of adequate length.

The general principles of tile hanging are similar to those of double lap roof tiling and the gauge is calculated in the same manner. The minimum lap recommended is 40 mm which would give a gauge of 112.5 mm using a standard 265 long tile.

A gauge dimension of 112.5 m is impracticable and therefore a gauge of 110 m would be usual. Typical details of top edge finishes, bottom edge finishes, corners and finishes at windows are shown in Figs. III.26 and III.27. It should be noted that if the structural backing is of timber framing a layer of impervious felt should be placed over the framing immediately underneath the battens to prevent any moisture which is blown in between the tiles from having adverse effects upon the structure. In this situation building paper is not considered to be a suitable substitute. The application of slates as a small unit hung cladding follows the principles outlined above for tile hanging.

TIMBER CLADDINGS

Timber claddings are usually in the form of moulded or shaped boards fixed to battens as either a horizontal or vertical cladding; typical examples are shown in Fig. III.28. Timber claddings will require regular maintenance to preserve their resistance to the elements. Softwoods are generally painted and will need repainting at intervals of three to five years according to the exposure. Hardwoods are sometimes treated with a preservative and left to bleach naturally; the preservative treatment needs to be carried out at two- to five-year intervals. Western red cedar is a very popular wood for timber cladding since it has a natural immunity to insect and fungi attack under normal conditions. It also has a pleasing natural red/brown colour which can be maintained if the timber is coated with a clear sealer such as polyurethane; however, it will bleach to a grey/white colour if exposed to the atmosphere. Plastic boards are a substitute for timber and are fixed in a similar manner.

Claddings to frame structures

The methods available to clad a frame structure are extensive and include panels of masonry constructed between the columns and beams, light infill panels of metal or timber, precast concrete panels and curtain walling which completely encloses the structure. A second-year course in construction technology confines itself

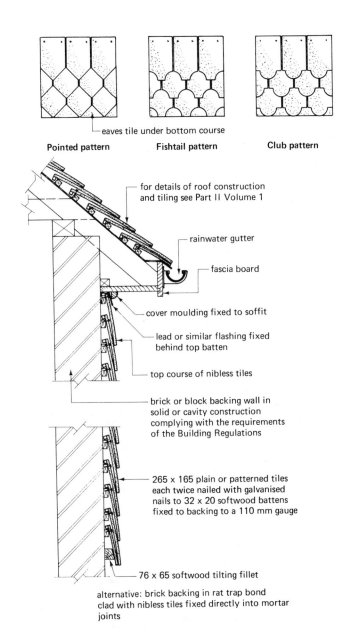

eaves tile under bottom course

Pointed pattern   **Fishtail pattern**   **Club pattern**

for details of roof construction
and tiling see Part II Volume 1

rainwater gutter

fascia board

cover moulding fixed to soffit

lead or similar flashing fixed
behind top batten

top course of nibless tiles

brick or block backing wall in
solid or cavity construction
complying with the requirements
of the Building Regulations

265 x 165 plain or patterned tiles
each twice nailed with galvanised
nails to 32 x 20 softwood battens
fixed to backing to a 110 mm gauge

76 x 65 softwood tilting fillet

alternative: brick backing in rat trap bond
clad with nibless tiles fixed directly into mortar
joints

Fig. III.26 Vertical tile hanging — typical details 1

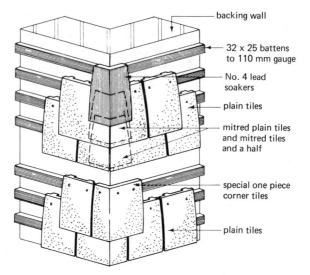

backing wall

32 x 25 battens
to 110 mm gauge

No. 4 lead
soakers

plain tiles

mitred plain tiles
and mitred tiles
and a half

special one piece
corner tiles

plain tiles

Alternative external angle treatments
(internal angles treatments similar)

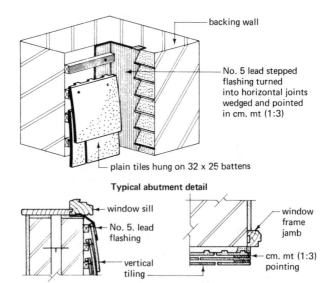

backing wall

No. 5 lead stepped
flashing turned
into horizontal joints
wedged and pointed
in cm. mt (1:3)

plain tiles hung on 32 x 25 battens

Typical abutment detail

window sill

No. 5. lead
flashing

vertical
tiling

window
frame
jamb

cm. mt (1:3)
pointing

Typical opening details

Fig. III.27 Vertical tile hanging — typical details 2

116

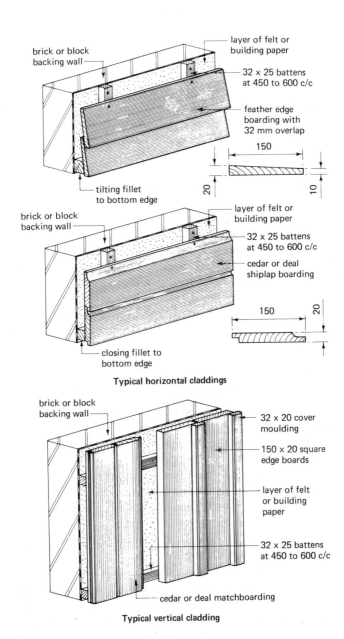

brick or block backing wall

layer of felt or building paper

32 × 25 battens at 450 to 600 c/c

feather edge boarding with 32 mm overlap

150

20

10

tilting fillet to bottom edge

brick or block backing wall

layer of felt or building paper

32 × 25 battens at 450 to 600 c/c

cedar or deal shiplap boarding

150

20

closing fillet to bottom edge

Typical horizontal claddings

brick or block backing wall

32 × 20 cover moulding

150 × 20 square edge boards

layer of felt or building paper

32 × 25 battens at 450 to 600 c/c

cedar or deal matchboarding

Typical vertical cladding

Fig. III.28 Timber wall claddings

117

to the study of panel walls and the facings which can be attached to them, the other forms of cladding are considered in courses concerned with advanced technology.

BRICK PANEL WALLS

These are non-load bearing walls which must fulfil the following requirements:

1. Adequate resistance to the elements.
2. Have sufficient strength to support their own self weight plus any attached finishes.
3. Strong enough to resist both positive and negative wind pressures.
4. Provide the required thermal and sound insulation.
5. Provide the required fire resistance.
6. Have adequate durability.

Brick panel walls are constructed in the same manner as ordinary solid or cavity walls and any openings for windows or doors are formed by traditional methods. The panels must be supported at each structural floor level and tied to the structure at the vertical edges. Projection of the panel in front of the structural members is permissible providing such overhangs do not impair the stability of the panel wall; acceptable limits are shown in Fig. III.29. The top edge of the panels should not be pinned rigidly to the frame since the effect of brick panel expansion together with frame shrinkage may cause cracking and failure of the brickwork. A compression joint should therefore be formed between the top edge of the panel and the underside of the framing member at each floor level (see Fig. III.29).

Two methods of tying the panel to the vertical structural members are in common use:

1. Butterfly wall tiles are cast into the column and built into the brick joints at four-course intervals.
2. Galvanised pressed steel dovetail slots are cast into the column and dovetail anchors are used to form the tie (see Fig. III.29).

The second method gives greater flexibility with the location and insertion of adequate ties but is higher in cost.

Facings to brick panel walls

Any panel wall must have an acceptable and durable finish; this can be achieved by using facing bricks with a neat pointed joint or by attaching to the face of a panel of common bricks a stone or similar cladding. Suitable materials are natural stone, artificial stone, reconstructed stone and precast concrete of small units up to 1 m² and with a thickness related to the density of the material. Dense materials such as slate and marble

118

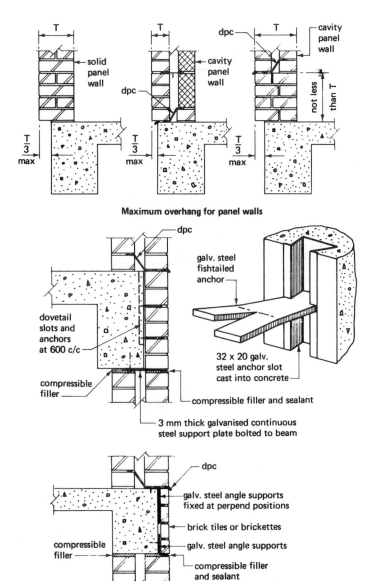

Fig. III.29 Brick panel walls

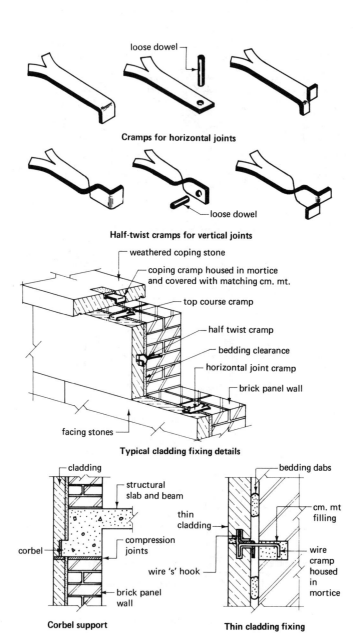

loose dowel

Cramps for horizontal joints

loose dowel

Half-twist cramps for vertical joints

weathered coping stone

coping cramp housed in mortice
and covered with matching cm. mt.

top course cramp

half twist cramp

bedding clearance

horizontal joint cramp

brick panel wall

facing stones

Typical cladding fixing details

cladding

structural
slab and beam

bedding dabs

thin
cladding

cm. mt
filling

compression
joints

corbel

wire 's' hook

wire
cramp
housed
in
mortice

brick panel
wall

Corbel support

Thin cladding fixing

Fig. III.30 Cladding fixings

120

need only be 40 mm thick, whereas the softer stones such as sandstone and limestone should be at least 75 mm thick.

Two major considerations must be taken into account when deciding on the method to be used to fix the facings to the brick backing:

1. Transferring the load to the structure.
2. Tying back the facing units.

The load of the facings can be transferred by using bonder stones or support corbels at each floor level, which should have a compression joint incorporated in the detail for the same reasons given above when considering brick panels (see Fig. III.30).

The tying back of the facings is carried out by various metal fixing devices called cramps which should be of a non-ferrous metal such as gunmetal, copper, phosphor bronze or stainless steel. To avoid the problem of corrosion caused by galvanic action between dissimilar metals a mixture of fixing materials should not be used. Typical examples of fixings and cramps for thick and thin facings are shown in Fig. III.30.

To provide for plumbing and alignment a bedding space of 12—15 mm should be left between the face of the brick panel and the back of the facing. Dense facings such as marble are usually bedded on a series of cement mortar dabs, whereas the more porous facings are usually placed against a solid bed which ensures that any saturation which occurs will be uniform over the entire face.

Part IV
Floors
and roofs

The function of any floor is to provide a level surface which is capable of supporting all the live and dead loads imposed. Reinforced concrete with its flexibility in design, good fire resistance and sound insulating properties is widely used for the construction of suspended floors for all types of buildings. The disadvantages of *in situ* concrete are:

1. Need for formwork.
2. Time taken for the concrete to cure before the formwork can be released for reuse and the floor available as a working area.
3. Very little is contributed by a large proportion of the concrete to the strength of the floor.

Floors composed of reinforced precast concrete units have been developed over the years to overcome some or all of the disadvantages of *in situ* reinforced concrete slab. To realise the full economy of any one particular precast flooring system the design of the floors should be within the span, width, loading and layout limitations of the units under consideration; coupled with the advantages of repetition.

CHOICE OF SYSTEM

Before any system of precast concrete flooring can be considered in detail the following factors must be taken into account:

1. Maximum span.
2. Nature of support.
3. Weight of units.

4. Thickness of units.
5. Thermal insulation properties.
6. Sound insulation properties.
7. Fire resistance of units.
8. Speed of construction.
9. Amount of temporary support required.

The systems available can be considered as either precast hollow floors or composite floors; further subdivision is possible by taking into account the amount of temporary support required during the construction period.

Precast hollow floors

Precast hollow floor units are available in a variety of sections such as box planks or beams; tee sections, I beam sections and channel sections (see Fig. IV.1). The economies which can be reasonably expected over the *in situ* floor are:
1. 50% reduction in the volume of concrete.
2. 25% reduction in the weight of reinforcement.
3. 10% reduction in size of foundations.

The units are cast in precision moulds, around inflatable formers or foamed plastic cores. The units are laid side by side with the edge joints being grouted together; a structural topping is-not required but the upper surface of the units are usually screeded to provide the correct surface for the applied finishes (see Fig. IV.1). Little or no propping is required during the construction period but usually some means of mechanical lifting is required to off load and position the units. Hollow units are normally the cheapest form of precast concrete suspended floor for simple straight spans with beam or wall supports up to a maximum span of 20.000 m. They are not considered suitable where heavy point loads are encountered unless a structural topping is used to spread the load over a suitable area.

The hollow beams or planks give a flat soffit which can be left in its natural state or be given a skim coat of plaster; the voids in the units can be used to house the services which are normally incorporated in the depth of the floor. The ribbed soffit of the channel and tee units can be masked by a suspended ceiling; again the voids created can be utilised to house the services. Special units are available with fixing inserts for suspended ceilings, service outlets and edges to openings.

Composite floors

These floors are a combination of precast units and *in situ* concrete. The precast units which are usually prestressed or reinforced with high yield steel bars are used to provide the strength of

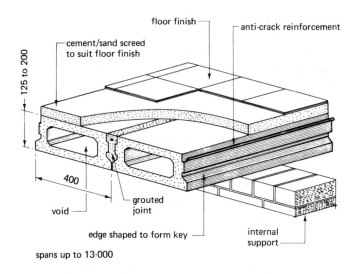

floor finish
anti-crack reinforcement
cement/sand screed
to suit floor finish
125 to 200
400
grouted
joint
void
edge shaped to form key
internal
support
spans up to 13·000

Typical hollow floor unit details

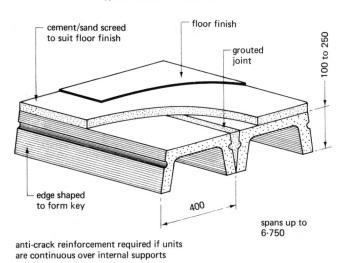

cement/sand screed
to suit floor finish
floor finish
grouted
joint
100 to 250
edge shaped
to form key
400
spans up to
6·750
anti-crack reinforcement required if units
are continuous over internal supports

Typical channel section floor unit details

Fig. IV.1 Precast concrete hollow floors

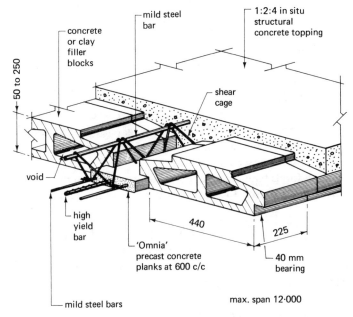

Typical composite floor using P.C.C. planks

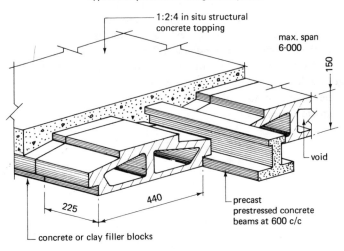

Typical composite floor using P.C.C. beams

Fig. IV.2 Composite floors

125

the floor with the smallest depth practicable and at the same time act as permanent formwork to the *in situ* topping which provides the compressive strength required. It is essential that an adequate bond is achieved between the two components — in most cases this is provided by the upper surface texture of the precast units; alternatively a mild steel fabric can be fixed over the units before the *in situ* topping is laid.

Composite floors generally take one of two forms:

1. Thin prestressed planks with a side key and covered with an *in situ* topping.

2. Reinforced or prestressed narrow beams which are placed at 600 mm centres and are bridged by concrete filler blocks, the whole combination being covered with *in situ* topping. Most of the beams used in this method have a shear reinforcing cage projecting from the precast beam section (see Fig. IV.2).

In both forms temporary support should be given to the precast units by props at 1.800 to 2.400 m centres until the *in situ* topping has cured.

COMPARISON OF SYSTEMS

Precast hollow floors are generally cheaper than composite, *in situ* concrete is not required and therefore the need for mixing plant and storage of materials is eliminated. The units are self centering, therefore temporary support is not required, the construction period is considerably shorter and generally the overall weight is less.

Composite floors will act in the same manner as an *in situ* floor and can therefore be designed for more complex loadings. The formation of cantilevers is easier with this system and support beams can be designed within the depth of the floor giving a flat soffit. Services can be housed within the structural *in situ* topping, or within the voids of the filler blocks. Like the precast hollow floor, composite floors are generally cheaper than a comparable *in situ* floor within the limitations of the system employed.

15

Hollow block
and waffle floors

Precast concrete suspended floors are generally considered to be for light to medium loadings spanning in one direction. Hollow block, or hollow pot floors as they are sometimes called, and waffle or honeycomb floors can be used as an alternative to the single spanning precast floor since they can be designed to carry heavier loadings. They are in fact ribbed floors consisting of closely spaced narrow and shallow beams giving an overall reduction in depth of the conventional reinforced concrete *in situ* beam and slab floor.

Hollow block floors

These are formed by laying over conventional floor soffit formwork a series of hollow lightweight clay blocks or pots in parallel rows with a space between these rows to form the ribs. The blocks act as permanent formwork giving a flat soffit suitable for plaster application and impart to the floor good thermal insulation and fire resistance. The ribs formed between the blocks can be reinforced to suit the loading conditions of the floor, thus providing flexibility of design (see Fig. IV.3). The main advantages of this system are its light weight, which is generally less than comparable floors of concrete construction, and its relatively low cost.

Waffle or honeycomb floors

Used mainly as an alternative to an *in situ* flat slab or a beam and slab suspended floor since it requires less concrete,

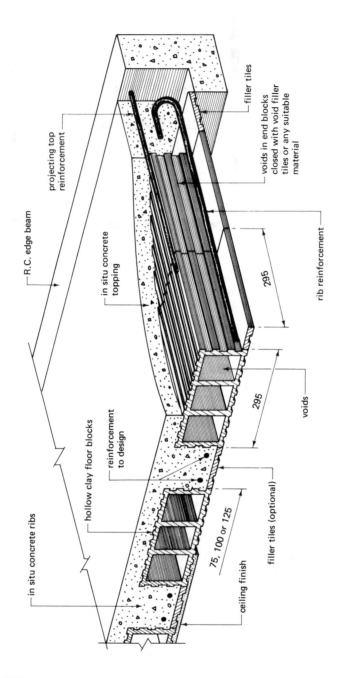

filler tiles

voids in end blocks
closed with void filler
tiles or any suitable
material

projecting top
reinforcement

R.C. edge beam

in situ concrete
topping

rib reinforcement

295

reinforcement
to design

hollow clay floor blocks

295

voids

in situ concrete ribs

75, 100 or 125

filler tiles (optional)

ceiling finish

Fig. IV.3 Hollow block floors

128

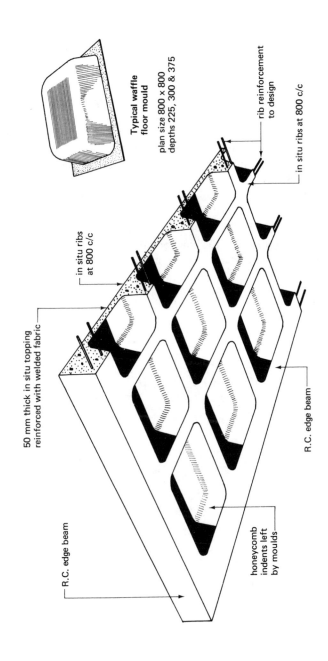

Typical waffle floor mould

plan size 800 × 800
depths 225, 300 & 375

rib reinforcement to design

in situ ribs at 800 c/c

in situ ribs at 800 c/c

50 mm thick in situ topping reinforced with welded fabric

R.C. edge beam

R.C. edge beam

honeycomb indents left by moulds

Fig. IV.4 Waffle or honeycomb floors

129

less reinforcement and can be used to reduce the number of beams and columns required with the resultant savings on foundations. The honeycomb pattern on the underside can add to the visual aspect of the ceiling by casting attractive shadow patterns.

The floor is cast over lightweight moulds or pans made of glass fibre, polypropylene or steel forming a two-directional ribbed floor (see Fig. IV.4). The moulds are very strong, lightweight and are capable of supporting all the normal loads encountered in building works. Support is reduced to the minimum since the moulds are arranged in parallel rows and span between the parallel lines of temporary supports.

The reinforcement in the ribs is laid in two directions to resist both longitudinal and transverse bending moments in the slab. Generally three mould depths are available but the overall depth can be increased by adding to the depth of the topping.

With all floors using an *in situ* topping it is possible to float the surface in preparation for the applied finishes, but this surface may suffer damage whilst being used by the following trades. It is therefore considered a better form of construction to allow for a floor screed to be applied to the *in situ* topping at a later stage in the contract prior to the fixing of the applied finish.

16

Steel roof trusses and coverings

The function of any roof is to provide a protective covering to the upper surface of the structure. By virtue of its position a roof is subjected to the elements to a greater degree than the walls, therefore the durability of the covering is of paramount importance. The roof structure must have sufficient strength to support its own weight, the load of the coverings together with any imposed loadings such as snow and wind pressures without collapse or excessive deflection.

Roofs can be considered as:

Short span: up to 7.000 m, generally of traditional timber construction with a flat or pitched profile. Flat roofs are usually covered with a flexible sheet material whereas pitched roofs are generally covered with small units such as tiles or slates.

Medium span: 7.000 to 24.000 m, except where reinforced concrete is used; the usual roof structure for a medium span is a truss or lattice of standard steel sections supporting a deformed sheeting such as corrugated asbestos cement or a structural decking system.

Long span: over 24.000 m; these roofs are generally designed by a specialist using girder, space deck or vaulting techniques and are beyond the scope of a basic technology course.

STEEL ROOF TRUSSES

This form of roof structure is used mainly for short and medium span single storey buildings intended for industrial or recreational use. A steel roof truss is a plane frame consisting of a series of rigid triangles composed of compression and tension members. The

131

compression members are called rafters and struts, whereas the tension members are termed ties. Standard mild steel angles complying with the recommendations of BS 4848 are usually employed as the structural members and these are connected together, where the centre lines converge, with flat shaped plates called gussets. They can be rivetted, bolted or welded together to form a rigid triangulated truss; typical arrangements are shown in Fig. IV.5. The internal arrangement of the struts and ties will be governed by the span. The principal or rafter is divided into a number of equal divisions which locates the intersection point for the centre line of the internal strut or tie.

Angle purlins are used longitudinally to connect the trusses together and provide the fixing medium for the roof covering. It is the type of covering chosen which will determine the purlin spacing and the pitch of the truss; ideally the purlins should be positioned over the strut or tie intersection points to avoid setting up local bending stresses in the rafters. Purlins are connected to cleats attached to the backs of the rafters; alternatively a zed section can be used, thus dispensing with the need for a fixing cleat. Steel roof trusses are positioned at 3.000 to 7.500 m centres and supported by capped universal columns or bolted to padstones bedded on to brick walls or piers. The main disadvantage of this form of roofing is the large and virtually unusable roof space. Other disadvantages are the frequent necessity of painting the members to inhibit rust and that the flanges of the angles provide an ideal ledge on which dust can accumulate. Typical steel roof truss details are shown in Fig. IV.6.

Suitable truss and girder arrangements can be fabricated from welded steel tubes which are lighter in weight, cleaner in appearance, have less surface area on which to collect dust and therefore less surface area to protect with paint.

Coverings

The basic requirements for covering materials to steel roof trusses are:

1. Sufficient strength to support imposed wind and snow loadings.
2. Resistance to the penetration of rain, wind and snow.
3. Low self weight, so that supporting members of an economic size can be used.
4. Reasonable standard of thermal insulation.
5. Acceptable fire resistance.
6. Durable to reduce the maintenance required during the anticipated life of the roof.

Most of the materials used for covering a steel roof structure have poor thermal insulation properties and therefore a combination of materials is

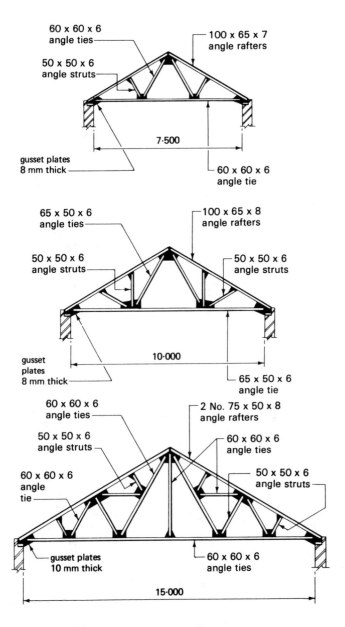

Fig. IV.5 Typical mild steel angle roof trusses

133

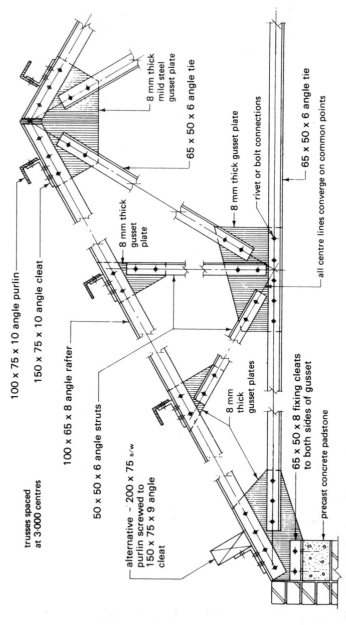

trusses spaced
at 3·000 centres

100 x 75 x 10 angle purlin

150 x 75 x 10 angle cleat

100 x 65 x 8 angle rafter

100 x 65 x 6 angle struts

50 x 50 x 6 angle struts

alternative ~ 200 x 75 s/w
purlin screwed to
150 x 75 x 9 angle
cleat

8 mm thick
mild steel
gusset plate

65 x 50 x 6 angle tie

8 mm thick
gusset
plate

8 mm thick gusset plate

rivet or bolt connections

65 x 50 x 6 angle tie

all centre lines converge on common points

8 mm thick
gusset plates

65 x 50 x 8 fixing cleats
to both sides of gusset

precast concrete padstone

Fig. IV.6 Typical medium span mild steel roof truss

134

required if heat loss, or gain, is to be controlled to satisfy legal requirements or merely to conserve the fuel required to heat the building.

Corrugated roofing materials, if correctly applied, will provide a covering which will exclude the rain and snow but will allow a small amount of wind penetration unless the end laps are sealed with 25 mm wide asbestos tape or a ribbon of mastic. These coverings are designed to support normal snow loads but are not usually strong enough to support the weight of operatives and therefore a crawling ladder or board should be used.

Owing to the poor thermal insulation value of these roofing materials there is a risk of condensation occurring on the internal surface of the sheets. This risk can be reduced by using a suitable lining at rafter level or by a ceiling at the lower tie level. Unless a vapour barrier is included on the warm side of the lining water vapour may pass through the lining and condense on the underside of the covering material, which could give rise to corrosion of the steel members. An alternative method is to form a 25 mm wide cavity between the lining and the covering.

GALVANISED CORRUGATED STEEL SHEETS
These sheets are often referred to as 'corrugated iron' and have been widely used for many years for small industrial and agricultural buildings; they can also be used as a cladding to post and rail fencing. They are generally made to the recommendations of BS 3083 which specifies the sizes, number of corrugations and the quality of the zinc coating or galvanising.

The pitch of the corrugations, which is the distance between centres of adjacent corrugations, is 75 mm with 7, 8, 9, or 10 corrugations per sheet width with lengths ranging from 1.500 to 3.600 m. A wide range of fittings for ridge, eaves and verge treatments are available. The sheets are secured to purlins with hook bolts, drive screws or nuts and bolts in a similar manner to that detailed for asbestos cement sheets in Fig. IV.9. The purlins are spaced at centres from 1.500 to 3.000 m according to the thickness of the sheeting being used. To form a weather-tight covering the sheets are lapped at their ends and sides according to the pitch and exposure conditions:

end laps: up to $20°$ pitch 150 mm minimum and sealed with a bituminous mastic;

side laps: formed on edge away from the prevailing wind with a 1½ corrugation lap for conditions of normal exposure and two corrugation lap for conditions of severe exposure.

When fixed, galvanised corrugated steel sheets form a roof which is cheap to construct, strong, rigid and non-porous. On exposure the

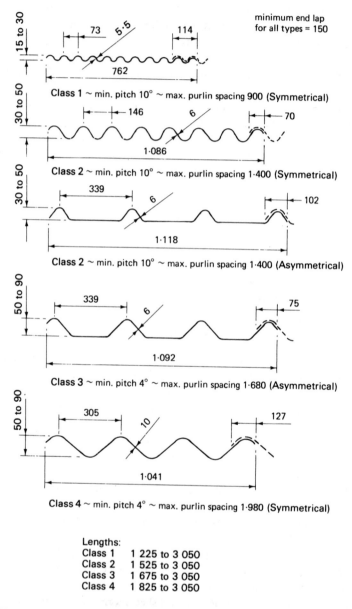

Fig. IV.7 Typical corrugated sheet profiles

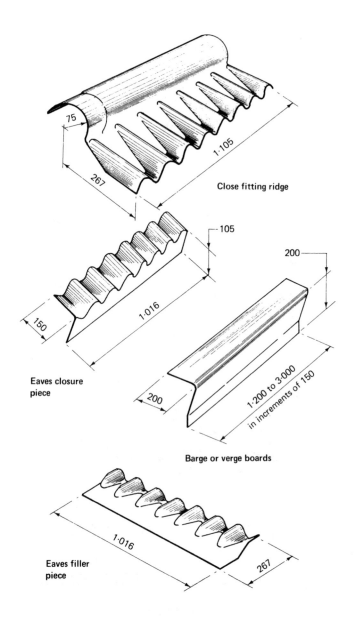

75

267

1·105

Close fitting ridge

105

200

150

1·016

Eaves closure
piece

1·200 to 3·000
in increments of 150

200

Barge or verge boards

1·016

267

Eaves filler
piece

Fig. IV.8 Typical fittings for Class 2 sheeting

137

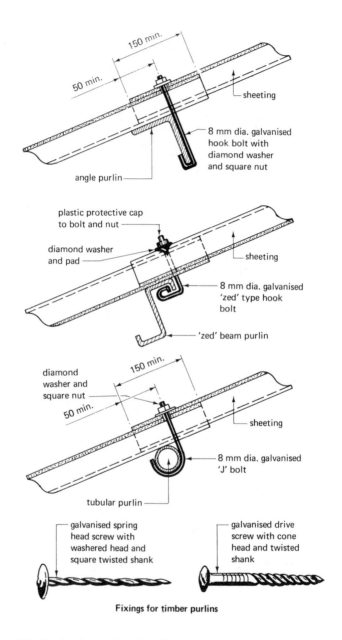

150 min.

50 min.

sheeting

8 mm dia. galvanised
hook bolt with
diamond washer
and square nut

angle purlin

plastic protective cap
to bolt and nut

diamond washer
and pad

sheeting

8 mm dia. galvanised
'zed' type hook
bolt

'zed' beam purlin

diamond
washer and
square nut

150 min.

50 min.

sheeting

8 mm dia. galvanised
'J' bolt

tubular purlin

galvanised spring
head screw with
washered head and
square twisted shank

galvanised drive
screw with cone
head and twisted
shank

Fixings for timber purlins

Fig. IV.9 Typical roof sheeting fixings

138

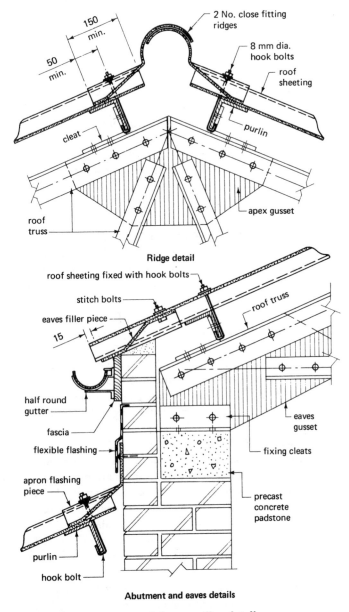

Ridge detail

2 No. close fitting ridges

8 mm dia. hook bolts

roof sheeting

150 min.

50 min.

cleat

purlin

roof truss

apex gusset

roof sheeting fixed with hook bolts

stitch bolts

eaves filler piece

15

roof truss

half round gutter

fascia

flexible flashing

apron flashing piece

purlin

hook bolt

eaves gusset

fixing cleats

precast concrete padstone

Abutment and eaves details

Fig. IV.10 Typical corrugated sheet roofing details

galvanised coating oxidises, forming a thin protective film which is easily broken down by acids in the atmosphere. To extend the life of the sheeting it should be regularly coated with paint containing a pigment of zinc dust, zinc oxide, calcium plumbate or zinc chromate. The use of these paints will eliminate the need for an application of mordant solution to provide a key. When laying new sheeting it is advisable to paint under the laps before fixing since the overlap is very vulnerable to early corrosion.

The main disadvantages of this form of roof covering are:

1. Poor thermal insulation properties – 8.6 W/m^2 °C which can be reduced by using a 12 mm insulation fibre board in conjunction with a 25 mm cavity to 1.7 W/m^2 °C.
2. Although a non-combustible material, galvanised corrugated steel sheets tend to buckle under typical fire conditions.
3. Inclined to be noisy during rain which produces a 'drumming' sound.

CORRUGATED ASBESTOS CEMENT SHEETS

This was the major covering material used for cladding steel roof structures and is made from asbestos fibres and cement in the approximate proportions of 1 : 7 together with a controlled amount of water. Corrugated asbestos cement sheets are produced to the recommendations of BS 690: Part 3 together with a wide range of fittings for the ridge, eaves and verge which are used in conjunction with the various profiles produced (see Figs. IV.7 and IV.8).

Concern with the health risk attached to the manufacture and use of asbestos-based products has led to the development and production of alternative fibre-based materials including profiles to match the corrugated asbestos cement sheets conforming to BS 690:Part 3.

FIBRE CEMENT PROFILED SHEETS

This alternative material to asbestos cement has been developed to meet the same technical specification with a similar low maintenance performance. Fibre cement sheets are made by combining natural and synthetic non-toxic fibres and fillers with Portland Cement and, unlike asbestos cement sheets which are rolled to form the required profile, these sheets are pressed over templates. The finished product has a natural grey colour but sheets with factory applied surface coatings are available. No British Standard yet exists for fibre cement sheeting but some products are supported by an Agrément Certificate, therefore for details of the properties, spanning ability, roof pitches and dimensions the individual manufacturer's data should be consulted.

The sheets and fittings are fixed through the crown of the corrugation using either shaped bolts to steel purlins or drive screws to timber purlins. At least six fixings should be used for each sheet and to ensure that a weather-tight seal is achieved at the fixing positions a suitable felt or lead pad with a diamond shaped curved washer can be used. Alternatively a plastic sealing washer can be employed (see Fig. IV.9). The sheets can be easily drilled for fixings which should be 2 mm larger in diameter than the fixing and sited at least 40 mm from the edge of the sheet. Side laps should be positioned away from the prevailing wind and end laps on low pitches should be sealed with a mastic or suitable preformed compressible strip.

The 'U' value of fibre cement sheets is high, generally about 6.0 W/m^2 K, therefore if a higher degree of thermal resistance is required it will be necessary to use a system of underlining sheets with an insulating material sandwiched between the profile and underlining sheet.

ALUMINIUM SHEETING

This form of roof covering is available in a corrugated or troughed profile usually conforming to the requirements of BS 4868. The sheets are normally made from an aluminium-manganese alloy resulting in a non-corrosive, non-combustible lightweight sheet (2.4 to 5.0 kg/m^2). Aluminium sheeting oxides on the surface to form a protective film upon exposure to the atmosphere and therefore protective treatments are not normally necessary. Fixings of copper or brass should not be used since the electrolytic action between the dissimilar materials could cause harmful corrosion, and where the sheets are in contact with steelwork the steel members should be painted with at least two coats of zinc chromate or bituminous paint.

The general application in design and construction of an aluminium covering is similar to that described and detailed for asbestos cement sheeting. A wide range of fittings are available and like the asbestos cement sheets can be fixed with hook bolts, bolts and clips or special shot fasteners. The sheets are intended for a 15° pitched roof with purlins at 1.200 m centres for the 75 mm corrugated profile and at 2.700 m centres for the trough profiles. Laps should be 1½ corrugations for the side lap or 45–57 mm for trough sheets with a 150 mm minimum end lap for all profiles.

17

Asphalt flat roofs

Flat roofs are often considered to be a simple form of construction, but unless correctly designed and constructed they can be an endless source of trouble. Flat roofs can have the advantage of providing an extra area to a building for the purposes of recreation and an additional viewpoint. Mastic asphalt provides an ideal covering material especially where foot traffic is required.

Mastic asphalt consists of an aggregate with a bituminous binder which is cast into blocks ready for reheating on site (see pages 174–5, Volume I). The blocks are heated in cauldrons or cookers to a temperature of over 200° C and are then transported in a liquid state in buckets for application to the roof deck by hand spreading. Once the melted asphalt has been removed from the source of heat it will cool and solidify rapidly, therefore the distance between the cauldron and the point of application should be kept to a minimum. Mastic asphalt can be applied to most types of rigid sub-structure and proprietary structural deckings. To prevent slight movements occurring between the sub-structure and the finish an isolating membrane should be laid over the base before the mastic asphalt is applied.

Roof work will often entail laying mastic asphalt to horizontal, sloping and vertical surfaces and these are defined as follows:
1. Horizontal surfaces – up to 10° pitch.
2. Sloping surfaces – between 10° and 45° pitch.
3. Vertical surfaces – over 45° pitch.

The thickness and number of coats required will depend on two factors:
1. Surface type.

2. Sub-structure or base.

Horizontal surfaces with any form of rigid base should have a two-coat application of mastic asphalt laid breaking the joint and built up to a minimum total thickness of 20 mm. An isolating membrane of black sheathing felt complying with BS 747 4A(i) should be used as an underlay laid loose with 50 mm lapped joints.

Vertical and sloping surfaces other than those with a timber base, require a three-coat application built up to a 20 mm total thickness without an isolating membrane.

Timber sub-structures with vertical and sloping surfaces should have a three-coat 20 mm mastic asphalt finish applied to expanded metal lathing complying with BS 1369 fixed at 150 mm centres over an isolating membrane.

If the mastic asphalt surface is intended for foot traffic the total thickness of the two-coat application should be increased to a minimum of 25 mm.

Roofs with a mastic asphalt finish can be laid to falls so that the run off of water is rapid and efficient. Puddles of water left on the roof surface will create small areas of different surface temperatures which could give rise to local differential thermal movements and cause cracking of the protective covering. Depressions will also provide points at which dust and rubbish can collect; it is therefore essential that suitable falls are provided to direct the water to an eaves gutter or specific outlet points. The falls should be formed on the base or supporting structure to a gradient of not less than 1 in 80 (see typical details on Figs. IV.11 and IV.12).

Alternatively the roof can be designed to act as a reservoir by a technique sometimes called 'ponding'. The principle is to retain on the roof surface a 'pond' of water some 150 mm deep by having the surface completely flat, high skirtings and outlets positioned 150 mm above the roof level. The main advantage is that differential temperatures are reduced to a minimum; the disadvantages are the need for a stronger supporting structure to carry the increased load, a three-coat 30 mm thick covering of mastic asphalt and the need to flood the roof in dry hot weather to prevent the pond completely evaporating away.

Thermal insulation can be provided by including into the design a dry insulation such as cork slabs, wood fibre boards and glass fibre boards. The insulation may be placed above or below the structural roof; if fixed over the roof structure it will reduce the temperature variations within the roof to a minimum and hence the risk of unacceptable thermal movements. An alternative method is to use a lightweight concrete screed to provide the falls and the thermal insulation. Suitable aggregates are furnace clinker, foamed blast furnace slag, pumice, expanded clay, sintered pulverised fuel

144

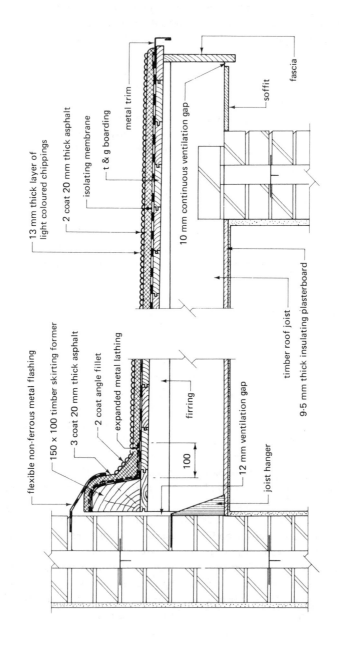

Fig. IV.11 Timber flat roof with mastic asphalt covering

13 mm thick layer of light coloured chippings

2 coat 20 mm thick asphalt

isolating membrane

t & g boarding

metal trim

10 mm continuous ventilation gap

soffit

fascia

timber roof joist

9·5 mm thick insulating plasterboard

flexible non-ferrous metal flashing

150 × 100 timber skirting former

3 coat 20 mm thick asphalt

2 coat angle fillet

expanded metal lathing

firring

100

12 mm ventilation gap

joist hanger

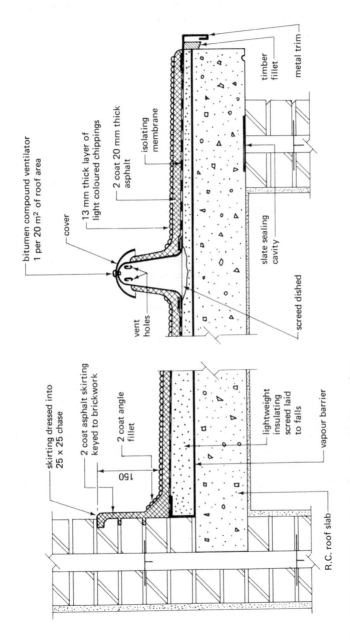

bitumen compound ventilator 1 per 20 m² of roof area

cover

13 mm thick layer of light coloured chippings

2 coat 20 mm thick asphalt

isolating membrane

timber fillet

metal trim

slate sealing cavity

screed dished

vent holes

skirting dressed into 25 x 25 chase

2 coat asphalt skirting keyed to brickwork

2 coat angle fillet

150

lightweight insulating screed laid to falls

vapour barrier

R.C. roof slab

Fig. IV.12 R.C. flat roof with mastic asphalt covering

145

ash, exfoliated vermiculite and expanded perlite. The thickness should not be less than 40 mm and it may be necessary to apply to the screed a 1:4 cement/sand topping to provide the necessary surface finish. Mix ratios of 1:8–10 are generally recommended for screeds of approximately 1 100 kg/m^3 density using foamed slag, sintered pulverised fuel ash and expanded clay, whereas a 1:5 mix is recommended for exfoliated vermiculite and expanded perlite aggregates giving a low density of under 640 kg/m^3. When mixing screeds containing porous aggregates high water/cement ratios are required to give workable mixes and therefore the screeds take a long time to dry out. Since the average rainfall in this country generally exceeds the rate of evaporation it is not always possible to ensure that the screed has dried out before the impermeable finish is applied. This entrapped moisture will tend to vaporise and cause an upward pressure during warm weather which could result in lifting or blistering of the mastic asphalt layer. To overcome this problem roof ventilators can be fixed to help relieve the pressure (see Fig. IV.12).

Moisture in the form of a vapour will tend to rise within the building and condense on the underside of the covering or within the thickness of the insulating material. If an insulating material becomes damp its efficiency will decrease and if composed of organic material it can decompose. To overcome this problem a vapour barrier of a suitable impermeable material such as polythene sheet or aluminium foil should be placed beneath the insulating layer. Care must be taken to see that all laps are complete and sealed and that the edges of the insulating material are similarly protected against the infiltration of moisture vapour.

The surface of a flat roof being fully exposed will gain heat by solar radiation and if insulated will be raised to a temperature in excess of the air or ambient temperature, since the transfer of heat to the inside of the building has been reduced by the insulating layer. The usual method employed to reduce the amount of solar heat gain of the covering is to cover the upper surface of the roof with light coloured chippings to act as a reflective finish. Suitable aggregates are white spar, calcined flint, white limestone or any light coloured granite of 12 mm size and embedded in a bitumen compound.

18

Lead-covered flat roofs

Lead as a building material has been used extensively for over 5 000 years and is obtained mainly from the mineral galena of which Australia, Canada, Mexico and the USA are the main producers. The raw material is mined, refined to a high degree of purity and then cast into bars or pigs which can be used to produce lead sheet, pipe and extruded products.

Lead is a durable and dense material (11 340 kg/m^3) of low strength but is very malleable and can be worked cold into complicated shapes without fracture. In common with other non-ferrous metals lead oxidises on exposure to the atmosphere and forms a thin protective film or coating over its surface. When in contact with other metals there is seldom any corrosion by electrolysis and therefore fixing is usually carried out by using durable copper nails.

For flat roofs milled lead sheet complying with the recommendations of BS 1178 is used. This sheet is supplied in rolls of a standard width of 2.400 m with lengths up to 12.000 m. For easy identification lead sheet carries a colour coding for each code number thus:

BS Code No.	Thickness (mm)	Colour
3	1.25	green
4	1.80	blue
5	2.24	red
6	2.50	black
7	3.15	white
8	3.55	orange

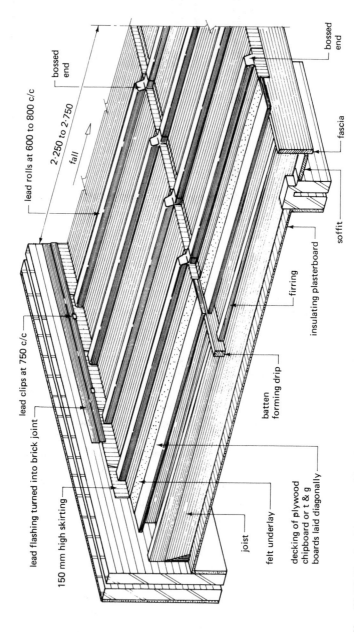

bossed end

bossed end

lead rolls at 600 to 800 c/c

2·250 to 2·750

fall

fascia

soffit

firring

insulating plasterboard

lead flashing turned into brick joint

lead clips at 750 c/c

150 mm high skirting

batten forming drip

joist

felt underlay

decking of plywood chipboard or t & g boards laid diagonally

Fig. IV.13 Typical layout of lead flat roof

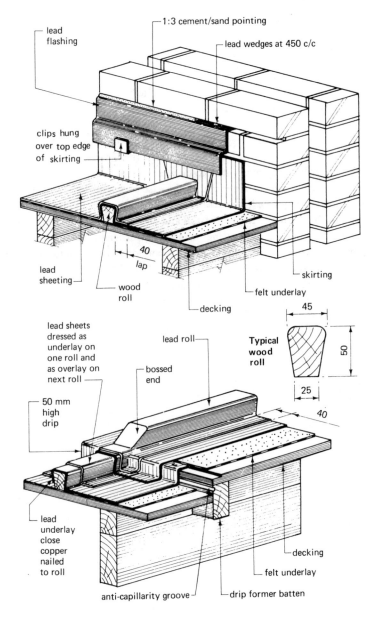

lead
flashing

1:3 cement/sand pointing

lead wedges at 450 c/c

clips hung
over top edge
of skirting

lead
sheeting

wood
roll

40
lap

decking

skirting

felt underlay

lead sheets
dressed as
underlay on
one roll and
as overlay on
next roll

lead roll

bossed
end

45

Typical
wood
roll

50

25

50 mm
high
drip

40

lead
underlay
close
copper
nailed
to roll

anti-capillarity groove

drip former batten

decking

felt underlay

Fig. IV.14 Lead flat roof details

The code number is derived from the former imperial notation of $5 \text{ lb/ft}^2 = \text{No. 5 lead}$.

The thickness or code number of lead sheet for any particular situation will depend upon the protection required against mechanical damage and the shape required. The following thickness can therefore be considered as a general guide for flat roofs:

1. Small areas without foot traffic No. 4 or 5.
2. Small areas with foot traffic No. 5, 6 or 7.
3. Large areas with or without foot traffic No. 5, 6 or 7.
4. Flashings No. 4 or 5.
5. Aprons No. 4 or 5.

Milled lead sheet may be used as a covering over timber or similar deckings and over smooth screeded surfaces. In all cases an underlay of felt or stout building paper should be used to reduce frictional resistances, decrease surface irregularities and in the case of a screeded surface isolate the lead from any free lime present which may cause corrosion. Provision must also be made for the expansion and contraction of the metal covering. This can be achieved by limiting the area and/or length of the sheets being used. It is recommended that the area of any one piece should not exceed 2.25 m^2 and the length should not exceed 2.500 m. Joints which can accommodate the anticipated thermal movements are in the form of rolls running parallel to the fall and drips at right angles to the fall positioned so that they can be cut economically from a standard sheet; for layout and construction details see Figs. IV.13 and IV.14.

19

Copper-covered flat roofs

Copper, like lead, has been used as a building material for many centuries. It is obtained from ore deposits which have to be mined, crushed and ground to a fine powder. Generally by a system of flotation the copper dust, which is about 4% of the actual rock mined, is separated from the waste materials and transferred to the smelting furnace and then cast into cakes of blister copper or thick slabs called anodes. The metal is now refined and formed into copper wire, strip, sheets and castings. Copper is also used to form alloys which are used in making components for the building industry. Common alloys are copper- and zinc-forming brass and copper- and tin-forming bronze.

Copper is a dense material (8 900 kg/m^3) which is highly ductile and malleable and can be cold worked into the required shape or profile. The metal hardens with cold working but its original dead soft temper can be restored by the application of heat with a blow-lamp or oxy-acetylene torch and quenching with water or by natural air cooling. If the dead soft temper is not maintained the hardened copper will be difficult to work and may fracture. On exposure to the atmosphere copper forms on its upper surface a natural protective film or patina, which varies in colour from green to black, making the copper beneath virtually inert.

For covering flat roofs rolled copper complying with the recommendations of BS 2870 is generally specified. Rolled copper is available in three forms:

Sheet: flat material of exact length, over 0.15 mm up to and including 10.0 mm thick and over 450 mm in width.

Strip: material over 0.15 mm up to and including 10.0 mm thick and any

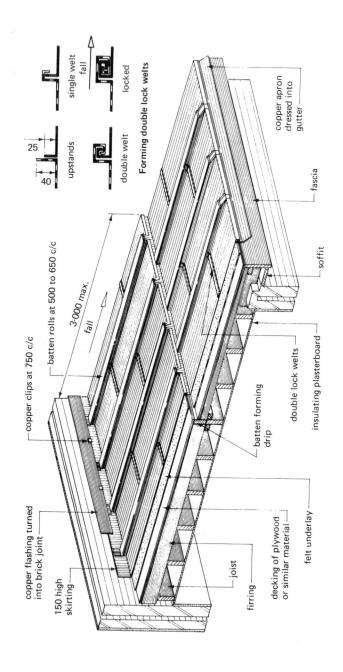

single welt

fall

locked

Forming double lock welts

25

40

upstands

double welt

copper apron dressed into gutter

fascia

soffit

batten rolls at 500 to 650 c/c

3·000 max.

fall

copper clips at 750 c/c

double lock welts

insulating plasterboard

batten forming drip

copper flashing turned into brick joint

150 high skirting

joist

firring

decking of plywood or similar material

felt underlay

Fig. IV.15 Typical layout of copper flat roof

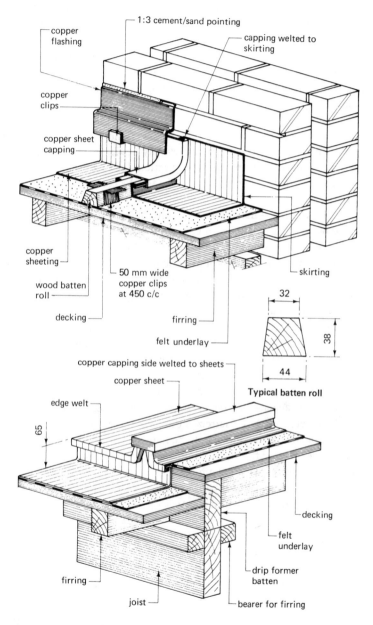

copper
flashing

1:3 cement/sand pointing

capping welted to
skirting

copper
clips

copper sheet
capping

copper
sheeting

wood batten
roll

decking

50 mm wide
copper clips
at 450 c/c

firring

felt underlay

skirting

32

38

44

Typical batten roll

copper capping side welted to sheets

copper sheet

edge welt

65

decking

felt
underlay

drip former
batten

firring

joist

bearer for firring

Fig. IV.16 Copper flat roof details

width, and generally not cut to length. It is usually supplied in coils but can be obtained flat or folded.

Foil: material 0.15 mm thick and under, of any width supplied flat or in a coil; because of its thickness foil has no practical application in the context of roof coverings.

In general copper strip is used for flashings and damp-proof courses, whereas sheet or strip can be considered for general roof covering application, widths of 600 and 750 mm, according to thickness, are used with a standard length of 1.800 m.

The copper sheet or strip is laid in bays between rolls or standing seams to allow for the expansion of the covering. Standing seams are not recommended for pitches under $5°$ since they may be flattened by foot traffic and become a source of moisture penetration due to a capillary action. The recommended maximum bay widths for the common sheet thicknesses used are:

    0.45 m thick — maximum bay width 600.

    0.60 m thick — maximum bay width 600.

    0.70 m thick — maximum bay width 750.

Transverse joints in the form of double lock welts should be used on all flat roofs. The welts should be sealed with a mastic or linseed oil before being folded together. Cross welts may be staggered or continuous across the roof when used in conjunction with batten rolls (see Fig. IV.15).

The substructure supporting the covering needs to be permanent, even, smooth and laid to the correct fall of not less than 1:60 (40 mm in 2.400 m). With a concrete structural roof a lightweight screed is normally laid to provide the correct surface and fall; if the screed has a sulphate content it will require sealing coats of bitumen to prevent any moisture present forming dilute acids which may react with the copper covering. The 50 x 25 dovetail battens inserted in the screed to provide the fixing medium for the batten rolls should be impregnated with a suitable wood preservative. Timber flat roofs should be wind tight and free from 'spring' and may be covered with 25 mm nominal tongued and grooved boarding laid with the fall or diagonally, or any other decking material which fulfils the general requirements given above.

On flat roofs drips at least 65 mm high are required at centres not exceeding 3 000 centres to increase the flow of rain water across the roof to the gutter or outlet. To lessen the wear on the copper covering as it expands and contracts a separating layer of felt (BS 747 Type 4A(ii)) should be incorporated into the design. The copper sheet should be secured with copper wire nails not less than 2.6 mm thick and at least 25 mm long; the batten rolls should be secured with well countersunk brass screws. Typical constructional details are shown in Fig. IV.16.

154

20
Timber stairs

Part V
Finishes and
fittings

The basic construction of straight flight timber stairs together with the requirements of the Building Regulations for private and public stairs are generally covered in the first year of study (see Volume I, Part III), whereas a second year course considers various layout arrangements and open tread or riserless stairs.

Layout arrangements

The introduction into a straight stair flight of landings or tapered steps will enable the designer to economise on the space required to accommodate the stairs. Landings can be quarter space giving a 90° turn or half space giving a 180° turn; for typical arrangements see Fig. V.1. The construction of the landing is similar to that of a timber upper floor except that with the reduced span joist depths can be reduced (see Fig. V.1). The landing can be incorporated in any position up the flight and if sited near the head may well provide sufficient headroom to enable a cupboard or cloakroom to be constructed below the stairs. A dog leg or string over string stair is economical in width since it will occupy a width less than two flights but this form has the disadvantage of a discontinuous handrail since this abuts to the underside of the return or upper flight.

Tapered steps

Prior to the introduction of the Building Regulations tapered steps or winders were frequently used by designers to use space economically since three treads occupied the area required for the conventional quarter space landing which is counted as one tread. These steps had the following disadvantages:

1. Hazard to the aged and very young because of the very small tread length at or near the newel post.
2. Difficult to carpet, requiring many folds or wasteful cutting.
3. Difficult to negotiate with furniture due to a rapid rise on the turn.
4. Have little or no aesthetic appeal.
5. Are expensive to construct.

With the introduction of the Building Regulations special attention has been given to the inclusion of tapered steps in Approved Document K which makes the use of tapered steps less of an economic proposition and more difficult to design (see Fig. V.2).

Open tread stairs

These are a contemporary form of stairs used in homes, shops and offices based on the simple form of access stair which has been used for many years in industrial premises. The concept is simplicity with the elimination of elaborate nosings, cappings and risers. The open tread or riserless stair must fully comply with Part K of the Building Regulations and in particular to Approved Document K which recommends a minimum tread overlap of 15 mm.

Three basic types of open tread stairs can be produced:

Closed string: which would be terminated at the floor and landing levels and fixed as for traditional stairs. The treads are tightly housed into the strings which are tied together with long steel tie bars under the first, last and every fourth tread. The nuts and washers can be housed into the strings and covered with timber inserts (see Fig. V.3).

Cut strings or carriages: these are used to support cantilever treads and can be worked from the solid or of laminated construction. The upper end of the carriage can be housed into the stairwell trimming member with possible additional support from metal brackets. The foot of the carriage is housed in a purpose made metal shoe or fixed with metal angle brackets (see Fig. V.4).

Mono-carriage: sometimes called a spine beam, employs a single central carriage with double cantilever treads. The carriage, which is by necessity large, is of laminated construction and very often of a tapered section to reduce the apparent bulky appearance. The foot of the carriage is secured with a purpose made metal shoe in conjunction with timber connectors (see Fig. V.5).

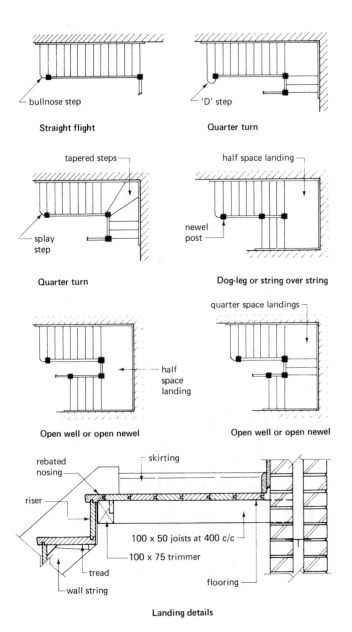

bullnose step

Straight flight

'D' step

Quarter turn

tapered steps

splay step

Quarter turn

half space landing

newel post

Dog-leg or string over string

half space landing

Open well or open newel

quarter space landings

Open well or open newel

rebated nosing

skirting

riser

100 x 50 joists at 400 c/c

100 x 75 trimmer

tread

wall string

flooring

Landing details

Fig. V.I Typical stair layouts

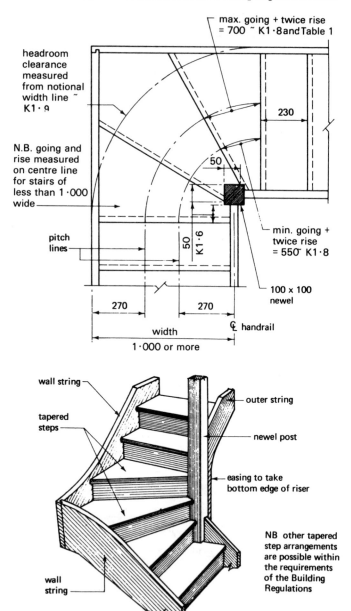

max. going + twice rise
= 700 ~ K1·8 and Table 1

headroom
clearance
measured
from notional
width line ~
K1·9

230

N.B. going and
rise measured
on centre line
for stairs of
less than 1·000
wide

50

min. going +
twice rise
= 550 K1·8

pitch
lines

50

50
K1·6

100 x 100
newel

270

270

℄ handrail

width
1·000 or more

wall string

outer string

tapered
steps

newel post

easing to take
bottom edge of riser

NB other tapered
step arrangements
are possible within
the requirements
of the Building
Regulations

wall
string

Fig. V.2 Tapered steps for private stairways

158

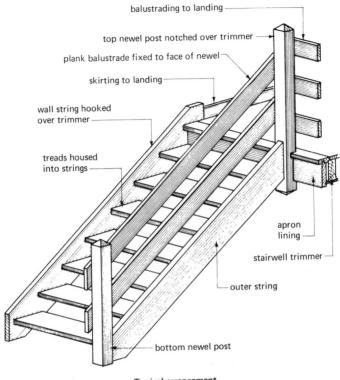

balustrading to landing

top newel post notched over trimmer

plank balustrade fixed to face of newel

skirting to landing

wall string hooked over trimmer

treads housed into strings

apron lining

stairwell trimmer

outer string

bottom newel post

Typical arrangement

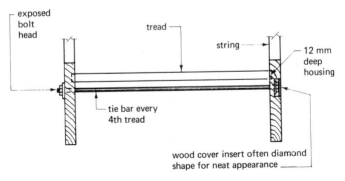

exposed bolt head

tread

string

12 mm deep housing

tie bar every 4th tread

wood cover insert often diamond shape for neat appearance

Alternative tie bar arrangements

Fig. V.3 Closed string open tread stairs

159

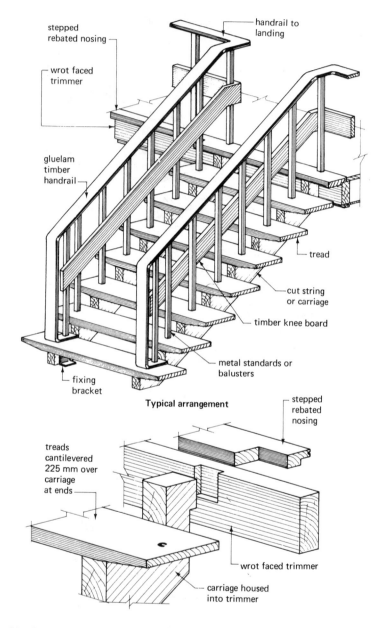

Typical arrangement

- stepped rebated nosing
- wrot faced trimmer
- handrail to landing
- gluelam timber handrail
- tread
- cut string or carriage
- timber knee board
- metal standards or balusters
- fixing bracket

treads cantilevered 225 mm over carriage at ends

- stepped rebated nosing
- wrot faced trimmer
- carriage housed into trimmer

Fig. V.4 Cut string open tread stairs

160

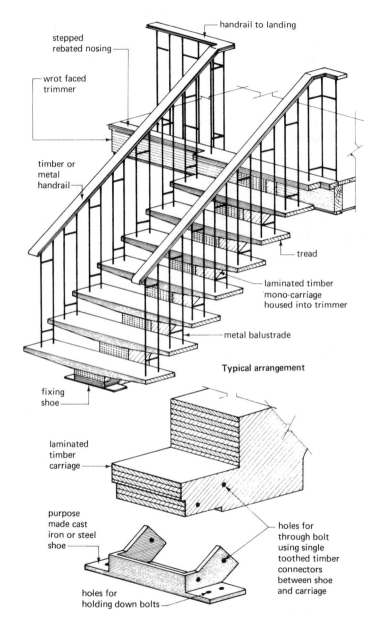

handrail to landing

stepped
rebated nosing

wrot faced
trimmer

timber or
metal
handrail

tread

laminated timber
mono-carriage
housed into trimmer

metal balustrade

fixing
shoe

Typical arrangement

laminated
timber
carriage

purpose
made cast
iron or steel
shoe

holes for
through bolt
using single
toothed timber
connectors
between shoe
and carriage

holes for
holding down bolts

Fig. V.5 Mono-carriage open tread stairs

161

Treads

These must be of adequate thickness since there are no risers to give extra support; usual thicknesses are 38 and 50 mm. To give a lighter appearance it is possible to taper the underside of the treads at their cantilever ends for a distance of 225–250 mm. This distance is based on the fact that the average person will ascend or descend a stairway at a distance of about 250 mm in from the handrail.

Balustrading

Together with the handrail balustrading provides both the visual and practical safety barrier to the side of the stair. Children present special design problems since they can and will explore any gap big enough to crawl through. BS 5395 for wood stairs recommends that the infill under handrails should have no openings which would permit the passage of a sphere 90 mm in diameter. Many variations of balustrading are possible ranging from simple newels with planks to elaborate metalwork of open design (see Figs. V.3, V.4, and V.5).

21

Simple reinforced concrete stairs

The functions of any stairway are:
1. To provide for the movement of people from one floor level to another.
2. To provide a suitable means of escape in case of fire.

A timber stairway will maintain its strength for a reasonable period during an outbreak of fire but will help to spread the fire thus increasing the hazards which could be encountered along an escape route. It is for this reason that the use of timber stairs are limited by Building Regulations B2 and B3 to certain domestic building types and stairs within shops which are not in a protective shaft and are not therefore part of the planned fire escape route.

Stairs, other than the exceptions given above, must therefore be constructed of non-combustible materials, but combustible materials are allowed to be used as finishes to the upper surface of the stairway or landing. Reinforced concrete stairs are non-combustible, strong and hard wearing. They may be constructed *in situ* or precast in sections ready for immediate installation and use when delivered to site. The general use of cranes on building sites has meant that many of the large flight arrangements, which in the past would have been cast *in situ*, can now be precast under factory controlled conditions.

Many variations of plan layout and spanning direction are possible but this study will be confined to the simple straight flight spanning from floor to floor or floor to intermediate landing. The designer will treat the stair as being an inclined slab spanning simply between the supports, the steps being treated as triangular loadings evenly distributed over the length.

163

Where intermediate landings are included in the design the basic plan is similar to the open well or newel timber stair. Difficulty is sometimes experienced with the intersection of the upper and lower flight soffits with the landing. One method of overcoming this problem is to have, in plan, the top and bottom steps out of line so that the soffit intersections form a straight line on the underside of the landing (see Fig. V.6). The calculations to determine the rise, going and number of steps is the same as those used for timber stairs; it should be noted that the maximum number of risers in each flight must not exceed sixteen. To achieve a greater tread length without increasing the actual horizontal going it is common to use a splayed riser face giving a 25 mm increase to the tread length.

The concrete specification is usually 1 : 2 : 4/20 mm aggregate with a cover of concrete over the reinforcement of 15 mm minimum or the bar diameter, whichever is the greater. This cover will give a one-hour fire resistance which is the minimum period implied by the Building Regulations. The thickness of concrete required is dependent on the loading and span but is not generally less than 100 mm or more than 150 mm measured across the waist which is the distance from the soffit to the intersection of tread and riser (see Fig. V.6).

Mild steel or high yield steel bars can be used to reinforce concrete stairs, the bars being lapped to starter bars at the ground floor and taken into the landing or floor support slab. The number, diameter and spacing of the main and distribution reinforcement must always be calculated for each stairway.

Handrails and balustrading must be constructed of a non-combustible material, continuous and to both sides if the width of the stairs exceeds 1.000 m. The overall height of the handrail up the stairs should be between 900 and 1 000 mm measured vertically and have a height above the floor of 1.100 m minimum. The capping can be of a combustible material such as plastic provided that it is fixed to or over a non-combustible core. Methods of securing balustrades and typical handrail details are shown in Fig. V.7.

A wide variety of finishes can be applied to the tread surface of the stairs. If the appearance is not of paramount importance, such as in a warehouse, a natural finish could be used but it would be advisable to trowel into the surface some carborundum dust to provide a hard-wearing non-slip surface. Alternatively, rubber or carborundum insert strips could be fixed or cast into the leading edges of the treads. Finishes such as PVC tiles, rubber tiles, and carpet mats are applied and fixed in the same manner as for floors. The soffits can be left as struck from the formwork and decorated or finished with a coat of spray plaster or a skim coat of finishing plaster.

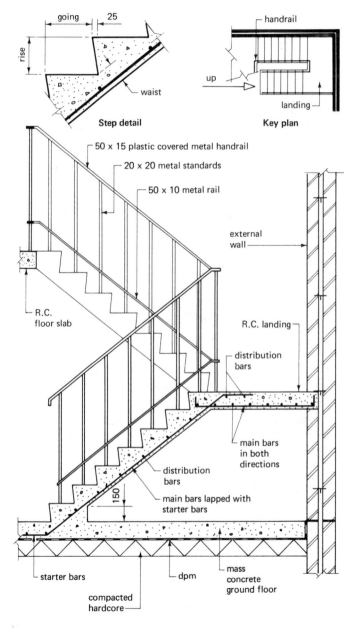

going   25

rise

waist

Step detail

handrail

up

landing

Key plan

50 x 15 plastic covered metal handrail

20 x 20 metal standards

50 x 10 metal rail

external wall

R.C. floor slab

R.C. landing

distribution bars

main bars in both directions

distribution bars

150

main bars lapped with starter bars

starter bars

dpm

mass concrete ground floor

compacted hardcore

Fig. V.6 Simple R.C. in situ **stairs**

165

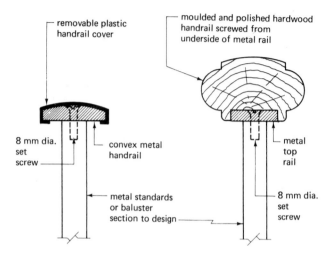

removable plastic
handrail cover

moulded and polished hardwood
handrail screwed from
underside of metal rail

8 mm dia.
set
screw

convex metal
handrail

metal
top
rail

8 mm dia.
set
screw

metal standards
or baluster
section to design

Typical handrails

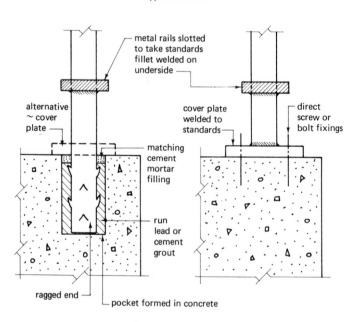

metal rails slotted
to take standards
fillet welded on
underside

alternative
~ cover
plate

cover plate
welded to
standards

direct
screw or
bolt fixings

matching
cement
mortar
filling

run
lead or
cement
grout

ragged end

pocket formed in concrete

Typical fixing methods

Fig. V.7 Handrails and balustrades

166

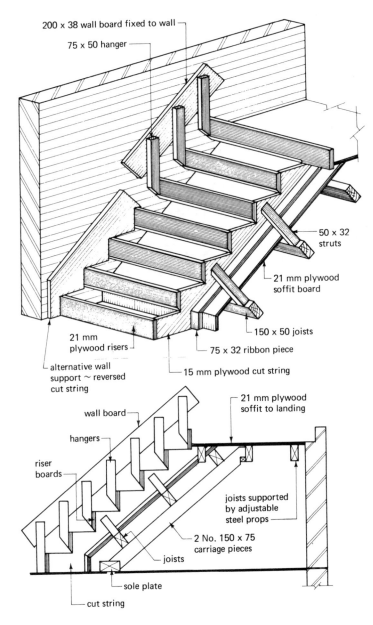

200 x 38 wall board fixed to wall

75 x 50 hanger

50 x 32 struts

21 mm plywood soffit board

150 x 50 joists

75 x 32 ribbon piece

21 mm plywood risers

15 mm plywood cut string

alternative wall support ~ reversed cut string

21 mm plywood soffit to landing

wall board

hangers

riser boards

joists supported by adjustable steel props

2 No. 150 x 75 carriage pieces

joists

sole plate

cut string

Fig. V.8 Typical formwork to R.C. in situ **stairs**

Formwork

The basic requirements are the same as for formwork to a framed structure. The stair profile is built off an adequately supported soffit of sheet material by using a cut string. Riser boards are used to form the leading face of the steps; these should have a splayed bottom edge to enable complete trowelling of the tread surfaces and to ensure that air is not trapped under the bottom edge of the riser board thus causing voids. If the stair abuts a vertical surface two methods can be considered to provide the abutment support for the riser boards; a reversed cut string or a wall board with hangers (see Fig. V.8). Wide stairs can have a reverse cut string as a central support to the riser boards to keep the thickness of these within an acceptable coat limit.

22

Simple precast concrete stairs

Precast concrete stairs can be designed and constructed to satisfy a number of different requirements. They can be a simple inclined slab, a cranked slab, an open riser stair or constructed from a series of precast steps built into, and if required cantilevered from, a structural wall.

The design considerations for the simple straight flight are the same as those for *in situ* stairs of comparable span, width and loading conditions. The fixing and support, however, require a different approach. Bearings for the ends of the flights must be provided at the floor or landing levels in the form of a haunch, rebate or bracket and continuity of reinforcement can be achieved by leaving projecting bars and slots in the floor into which they can be grouted (see Fig. V.9).

Ideally the delivery of precast stairs should be arranged so that they can be lifted, positioned and fixed direct from the delivery vehicle, thus avoiding double handling. Precast components are usually designed for two conditions:

1. Lifting and transporting.
2. Final fixed condition.

It is essential that the flights are lifted from the correct lifting points, which may be in the form of loops or hooks projecting from or recessed into the concrete member, if damage by introducing unacceptable stresses during lifting are to be avoided.

Balustrade and handrail requirements together with the various methods of fixing are as described for *in situ* reinforced concrete stairs. Any tread

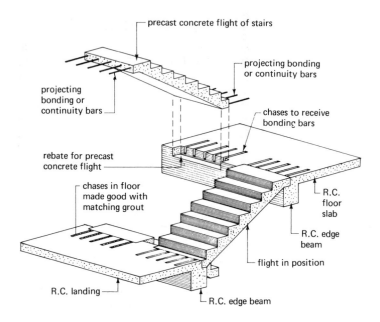

precast concrete flight of stairs

projecting bonding or continuity bars

projecting bonding or continuity bars

chases to receive bonding bars

rebate for precast concrete flight

chases in floor made good with matching grout

R.C. floor slab

R.C. edge beam

flight in position

R.C. landing

R.C. edge beam

Simple precast concrete stairs

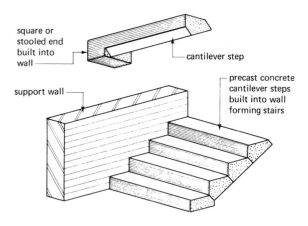

square or stooled end built into wall

cantilever step

support wall

precast concrete cantilever steps built into wall forming stairs

Simple precast concrete steps

Fig. V.9 Precast concrete stairs and steps

170

finish which is acceptable for an *in situ* stair will also be suitable for the precast alternative.

The use of precast concrete steps to form a stairway are limited to situations such as short flights between changes in floor level and external stairs to basements and areas. They rely on the load bearing wall for support and if cantilevered on the downward load of the wall to provide the necessary reaction. The support wall has to provide this necessary load and strength, and at the same time it has to be bonded or cut around the stooled end(s) of the steps. It is for these reasons that the application of precast concrete steps are restricted.

23
Windows

The function of a window is to allow the entry of light and air into the interior of the building. A window is essentially a frame of timber or metal set in an opening to which separate opening frames can be attached to provide the means of ventilation.

The main types of windows used are:

1. Casement windows — see Part III, Volume I.
2. Double hung or sliding sashes.
3. Pivoted casements.

BUILDING REGULATIONS 1985

The provision of windows as a means of ventilation is not required by Part F of the Building Regulations since the interior of any building can be adequately ventilated by mechanical means. Similarly the provision of natural daylight by means of windows is not required by Part F since any part of a building can be adequately illuminated by the provision of designed artificial lighting. It must be recognised however that a window not only provides daylight and ventilation but also a visual contact with the external surroundings, which is considered to be psychologically advantageous to the occupants. Part F of the Building Regulations which covers means of ventilation applies only to dwellings, rooms containing sanitary conveniences, bathrooms and buildings containing dwellings.

Building Regulation F1 requires that an adequate supply of air is provided for persons in the building. Approved Document F gives recommendations of ways of achieving this objective. The basic recommendations are:

1. Total area of ventilating opening(s) shall be not less than one-twentieth of the floor area of the room(s) it serves and one-fiftieth of the floor area for common spaces in buildings containing dwellings.
2. In habitable rooms some part of the ventilation opening(s) should be at least 1750mm above floor level.
3. A door opening directly to the external air can be classed as a ventilation opening.
4. Background ventilation is to be provided by ventilation opening(s) having a total area of not less than $4000mm^2$ by a means which is controllable, secure and sited to avoid undue draughts.

If a form of mechanical ventilation is to be used Approved Document F recommends the following:

1. Kitchens — intermittent rapid ventilation of not less than 60 litres per second or if incorporated in a cooker hood not less than 30 litres per second, both with suitable background ventilation.
2. Bathrooms — intermittent minimum extraction rate of 15 litres per second.
3. Sanitary Accommodation — intermittent minimum extraction rate of three air-changes per hour with a fifteen minute minimum over-run.
4. Common Spaces — minimum extraction rate of one air-change per hour.

Approved Document F also gives guidance as to the ventilation of a habitable room facing a wall which is within 15.000 m of the ventilation opening. Recommendations for both open and closed courts are given (see Fig. V.10).

It should be noted that apart from providing ventilation and natural daylight windows also contribute to heat loss from within the building. Therefore in the context of conservation of fuel and power the maximum aggregate area of windows and rooflights and/or the maximum calculated rate of heat loss are controlled by Part L of the Building Regulations. The requirements under Part L cover all types of buildings and are not, unlike Part F, concerned essentially with dwellings — see Chapter 26 on page 194.

Bay windows

Any window which projects in front of the main wall line is considered to be a bay window; various names are, however, given to various plan layouts (see Fig. V.11). Bay windows can

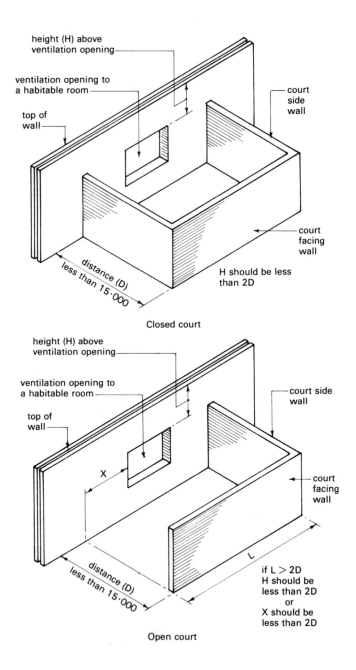

Fig. V.10 Ventilation openings facing courts

174

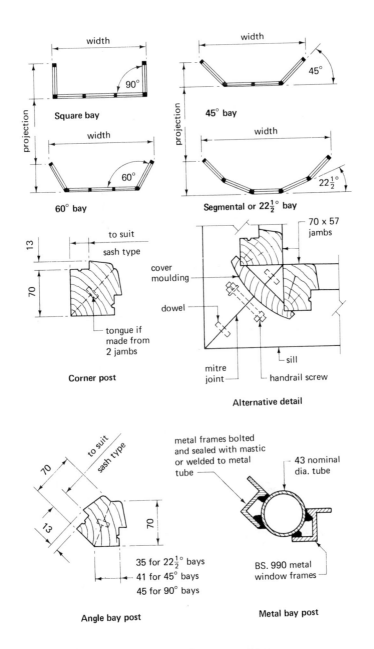

Fig. V.11 Bay window types and corner posts

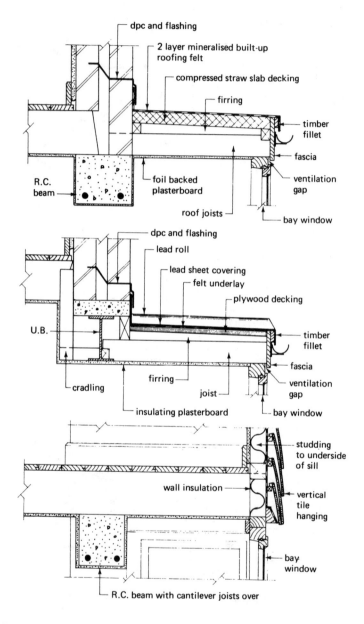

Fig. V.12 Bay window roofs and 2-storey bays

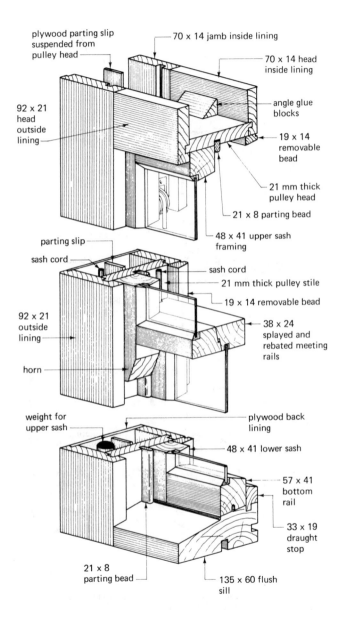

plywood parting slip suspended from pulley head

70 x 14 jamb inside lining

70 x 14 head inside lining

92 x 21 head outside lining

angle glue blocks

19 x 14 removable bead

21 mm thick pulley head

21 x 8 parting bead

48 x 41 upper sash framing

parting slip

sash cord

sash cord

21 mm thick pulley stile

19 x 14 removable bead

92 x 21 outside lining

38 x 24 splayed and rebated meeting rails

horn

weight for upper sash

plywood back lining

48 x 41 lower sash

57 x 41 bottom rail

33 x 19 draught stop

21 x 8 parting bead

135 x 60 flush sill

Fig. V.13 Double hung weight-balanced sash windows

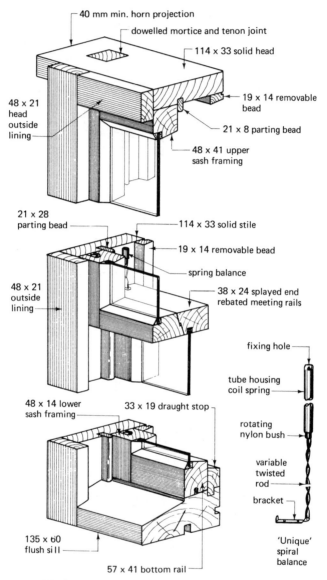

40 mm min. horn projection

dowelled mortice and tenon joint

114 x 33 solid head

48 x 21 head outside lining

19 x 14 removable bead

21 x 8 parting bead

48 x 41 upper sash framing

21 x 28 parting bead

114 x 33 solid stile

19 x 14 removable bead

spring balance

48 x 21 outside lining

38 x 24 splayed end rebated meeting rails

fixing hole

tube housing coil spring

rotating nylon bush

variable twisted rod

bracket

48 x 14 lower sash framing

33 x 19 draught stop

'Unique' spiral balance

135 x 60 flush sill

57 x 41 bottom rail

NB if 114 x 60 solid stiles are used balances can be housed in grooves within the stile thickness

Fig. V.14 Double hung spring-balanced sash windows

178

be constructed of timber, and/or metal and designed with casement or sliding sashes; the main difference in detail is the corner post, which can be made from the solid, jointed or masked in the case of timber and tubular for metal windows (see Fig. V.11).

The bay window can be applied to one floor only or continued over several storeys. Any roof treatment can be used to cover in the projection and weather seal it to the main wall (see Fig. V.12). No minimum head-room heights for bay windows or habitable rooms is given in the Building Regulations but 2.000 in bay windows and 2.300 in rooms would be considered reasonable. A bay window which occurs only on upper storeys is generally called an oriel window.

Double hung sash windows

These windows are sometimes called vertical sliding sash windows and consist of two sashes sliding vertically over one another. They are costly to construct but are considered to be more stable than side hung sashes and have a better control over the size of ventilation opening thus reducing the possibility of draughts.

In timber two methods of suspension are possible:
1. Weight balanced type.
2. Spring balanced type.

The former is the older method in which the counter balance weights suspended by cords are housed in a boxed framed jamb or mullion and have been generally superseded by the metal spring balance which uses a solid frame and needs less maintenance (see Figs. V.13 and V.14).

Double hung sashes in metal are supported and controlled by spring balances or by friction devices but the basic principles remain the same.

Pivot windows

The basic construction of the frame and sash is similar to that of a standard casement frame and sash. The sash can be arranged to pivot horizontally or vertically on friction pivots housed in the jambs or in the cill and head. These windows give good adjustment for ventilation purposes and in the main both faces of the glazing can be cleaned from the inside of the building.

24

Rooflights in pitched roofs

Rooflights can be included in the design of a pitched roof to provide daylight and ventilation to rooms within the roof space or to supplement the daylight from windows in the walls of medium and large span single storey buildings.

In domestic work a rooflight generally takes one of two forms — the dormer window or skylight. A dormer window has a vertical sash and therefore projects from the main roof, the cheeks or sides can be clad with a sheet material such as lead or tile hanging and the roof can be pitched or flat of traditional construction (see Fig. V.15).

A skylight is fixed in a trimmed opening and follows approximately the pitch of the roof. It can be constructed as an opening or dead light (see Fig. V.16). In common with all rooflights in pitched roofs, making the junctions water and weathertight present the greatest problems, and careful attention must be given to the detail and workmanship involved in the construction of dormer windows and rooflights.

Roofs of the type used on medium span industrial buildings with coverings such as corrugated fibre cement sheeting supported by purlins and steel roof trusses require a different treatment. The amount of useful daylight entering the interior of such a building from windows fixed in the external walls will depend upon:

1. The size of windows.
2. The height of window above the floor level.
3. The span of building.

Generally the maximum distance useful daylight will penetrate is approximately 10.000 m, over this distance artificial lights or roof lights

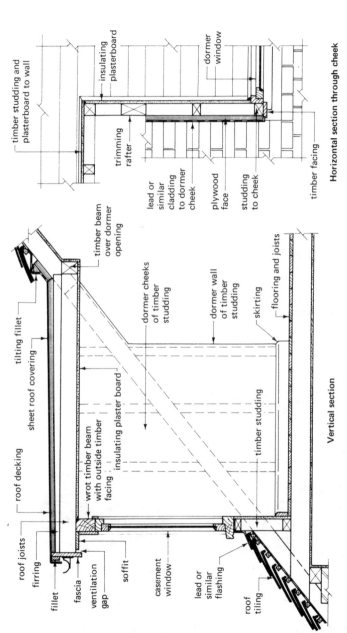

insulating plasterboard

dormer window

Horizontal section through cheek

timber studding and plasterboard to wall

trimming rafter

lead or similar cladding to dormer cheek

plywood face

studding to cheek

timber facing

timber beam over dormer opening

dormer cheeks of timber studding

dormer wall of timber studding

skirting

flooring and joists

tilting fillet

sheet roof covering

insulating plaster board

roof decking

wrot timber beam with outside timber facing

timber studding

Vertical section

roof joists

firring

fillet

fascia

ventilation gap

soffit

casement window

lead or similar flashing

roof tiling

Fig. V.15 Typical flat roof dormer window details

181

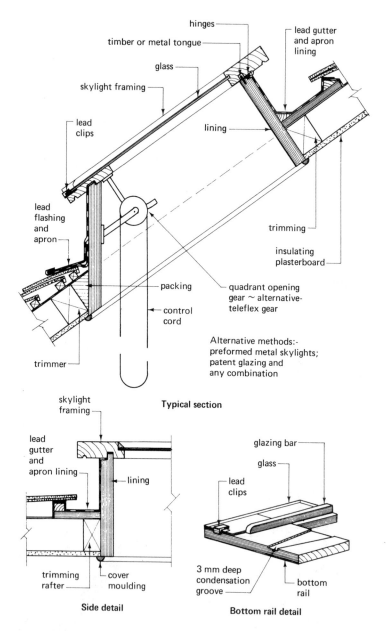

Typical section

Alternative methods:-
preformed metal skylights;
patent glazing and
any combination

Side detail

Bottom rail detail

Fig. V.16 Typical timber opening skylight

182

will be required during the daylight period. Three methods are available for the provision of roof lights in profiled sheeted roofs.

Special rooflight units of fibre cement consisting of an upstand kerb surmounted by either a fixed or opening glazed sash can be fixed instead of a standard profiled sheet. These units are useful where the design calls for a series of small isolated glazed rooflights to supplement the natural daylight. An alternative is to use translucent profiled sheets which are of the same size and profile as the main roof covering. In selecting the type of sheet to be used the requirements of Part B of the Building Regulations must be considered. Approved Document B (A3) deals specifically with the fire risks of roof coverings and refers to the designations defined in BS 476 : Part 3. These designations consist of two letters, the first letter represents the time of penetration when subjected to external fire and the second letter is the distance of spread of flame along the external surface. Each group of designations has four letters and in both cases the letter A has the highest resistance. Specimens used in the BS 476 test are either tested for use on a flat surface or a sloping surface and therefore the material designation is preceded by either EXT. F or EXT. S.

Most of the translucent profiled sheets have a high light transmission, are light in weight and can be fixed in the same manner as the general roof covering. It is advisable to weather seal all lapping edges of profiled rooflights with asbestos tape or mastic to accommodate the variations in thickness and expansion rate of the adjacent materials. Typical examples are:

Polyester glass fibre sheets: made from polyester resins reinforced with glass fibre and nylon to the recommendations of BS 4154. These sheets can be of natural colour or tinted and are made to suit most corrugated asbestos cement and metal profiles. Typical designations are EXT. S.AA for self-extinguishing sheets and EXT. S.AB for general-purpose sheets.

Wire reinforced PVC sheets: made from unplasticised PVC reinforced with a fine wire mesh to give a high resistance to shattering by impact. Designation is EXT. S.AA and they can therefore be used for all roofing applications. Profiles are generally limited to Class 1 and 2 defined in BS 690 :Part 3.

PVC sheets: made from heavy gauge clear unplasticised rigid PVC to the recommendations of BS 4203 are classified as self-extinguishing when tested in accordance with method 508A of BS 2782 and may be used on the roof of a building provided that part of the roof is at least 6.000 m from any boundary. If that part of the roof is less than

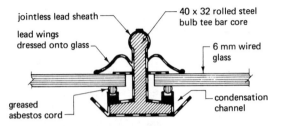

jointless lead sheath

40 x 32 rolled steel
bulb tee bar core

lead wings
dressed onto glass

6 mm wired
glass

greased
asbestos cord

condensation
channel

Crittall-Hope lead-clothed steel bar

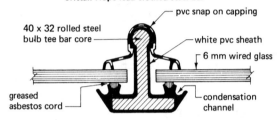

pvc snap on capping

40 x 32 rolled steel
bulb tee bar core

white pvc sheath

6 mm wired glass

greased
asbestos cord

condensation
channel

Crittall-Hope polyclad bar

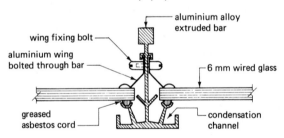

aluminium alloy
extruded bar

wing fixing bolt

aluminium wing
bolted through bar

6 mm wired glass

greased
asbestos cord

condensation
channel

British Challenge aluminium bar

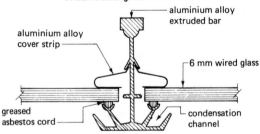

aluminium alloy
extruded bar

aluminium alloy
cover strip

6 mm wired glass

greased
asbestos cord

condensation
channel

Heywood Williams 'Aluminex' bar

Fig. V.17 Typical patent glazing bar sections

184

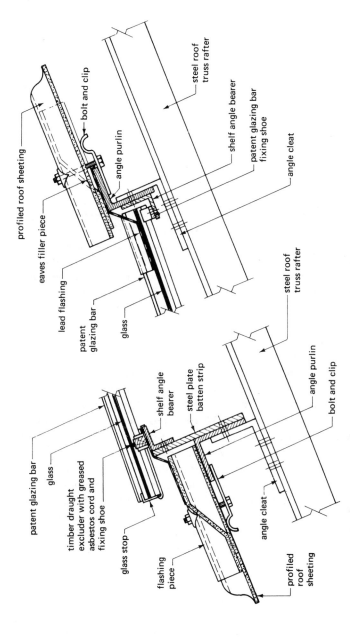

profiled roof sheeting

bolt and clip

angle purlin

steel roof truss rafter

shelf angle bearer

patent glazing bar fixing shoe

angle cleat

eaves filler piece

lead flashing

patent glazing bar

glass

steel roof truss rafter

angle purlin

bolt and clip

patent glazing bar

glass

timber draught excluder with greased asbestos cord and fixing shoe

glass stop

shelf angle bearer

steel plate batten strip

flashing piece

angle cleat

profiled roof sheeting

Fig. V.18 Patent glazing and profiled roof covering connection details

185

6.000 m from any boundary and covers a garage, conservatory or out-house with a floor area of less than 40 m² PVC sheets can be used without restriction. Table 6.3 of Approved Document B defines the use of these sheets for the roof covering of a canopy over a balcony, veranda, open carport, covered way or detached swimming pool.

As an alternative to profiled rooflights in isolated areas continuous rooflights can be incorporated into a corrugated or similar roof covering by using flat wired glass and patent glazing bars. The bars are fixed to the purlins and spaced at 600 mm centres to carry either single or double glazing. The bars are available as a steel bar sheathed in lead or PVC or as an aluminium alloy extrusion (see Fig. V.17). Many sections with different glass securing techniques are manufactured under the patents granted to the producers but all have the same basic principles. The bar is generally an inverted 'T' section, the flange providing the bearing for the glass and the stem depth giving the bar its spanning properties. Other standard components are fixing shoes, glass weathering springs or clips and glass stops at the bottom end of the bar (see Fig. V.17).

Since the glass and the glazing bar are straight they cannot simply replace a standard profiled sheet, they must be fixed below the general covering at the upper end and above the covering at the lower end to enable the rainwater to discharge on to the general roof surface. Great care must be taken with this detail and with the quality of workmanship on site if a durable and satisfactory junction is to be made. Typical details are shown in Fig. V.18.

The total amount of glazing to be used in any situation involves design appreciation beyond the scope of this volume, but a common rule of thumb method is to use an area of glazing equal to 10% of the total roof area. The glass specified is usually a wired glass of suitable thickness for the area of pane being used. Wired glass is selected so that it will give the best protection should an outbreak of fire occur; the splinters caused by the heat cracking the glass will adhere to the wire mesh and not shatter on to the floor below. As with timber skylights, provision should be made to collect the condensation which can occur on the underside of the glazing to prevent the annoyance of droplets of water falling to the floor below. Most patent glazing bars for single glazing have condensation channels attached to the edges of the flange which directs the collected condensa-tion to the upper surface of the roof below the glazing line (see Figs. V.17 and V.18).

25

Sound insulation

Part VI
Insulation

THE DEFINITION OF SOUND

Anything that can be heard is a sound, whether it is made by conversation, machinery, or walking on a hard surface. All sounds are produced by a vibrating object which moves rapidly to and fro causing movement of the tiny particles of air surrounding the vibrating source. The displaced air particles collide with adjacent particles setting them in motion and in unison with the vibrating object. Air particles move only to and fro but the sound wave produced travels through the air until at some distance from the source the movement of the particles is so slight that the sound produced is inaudible.

For a sound to be transmitted over a distance a substance, called the sound medium, is required. It can be shown by experiments that sound cannot travel through a vacuum but it can be transmitted through solids and liquids.

Sounds can differ in two important ways, by loudness and by pitch. The loudness of a sound depends on the distance through which the vibrating object moves to and fro as it vibrates; the greater the movement the louder the sound. The loudness with which a sound is heard depends upon how far away from the source the receiver or ear is. The unit of subjective loudness is a phon whilst the objective unit is called a decibel. Although the loudness of a sound will vary with the frequency of the note for practical building purposes, the phon and the decibel are considered to be equal over the range of 0 phons, the threshold of hearing, to 130 phons the threshold of painful hearing.

The pitch of a sound depends on the rate at which the vibrating object oscillates. The number of vibrations in a second is called the frequency and the higher the frequency the higher the pitch. The lowest pitched note that the human ear can hear has a frequency of approximately 16 hertz whereas the highest pitched note which can be heard by the human ear has a frequency of approximately 20 000 hertz or cycles per second.

When a sound is produced within a building three reactions can occur:

1. The pressure or sound waves can come into contact with the walls, floor and ceiling and be reflected back into the building.
2. Some of the sound can be absorbed by these surfaces and/or the furnishes. It must be noted that sound absorption normally only benefits the occupants of the room in which the sound is generated since its function is to reduce the amount of reflected sound.
3. The sound waves upon reaching the walls, floor and ceiling can set these members vibrating in unison and thus transmit the sound to adjacent rooms.

It must also be noted that sounds can enter a building from external sources such as traffic and low flying aircraft (see Fig. VI.1).

Sounds may be defined as either impact sounds, caused by direct contact with the structure such as footsteps and hammering on walls which will set that part of the structure vibrating, or they can be termed airborne sounds, such as the conversation or radio which sets the structure vibrating only when the sound waves emitted from the source reach the structural enclosure.

A noise can be defined as any undesired sound and may have any one of the following four effects on man:

1. Annoyance.
2. Disturbance of sleep.
3. Interfere with the ability to hold a normal conversation.
4. Damage his hearing.

It is difficult to measure annoyance since it is a subjective attitude and will depend upon the mental and physical well being of the listener, together with the experience of being subjected to such types of noise. Damage to hearing can be caused by a sudden noise such as a loud explosion or by gradual damage resulting from continual noise over a period of years.

The solution to noise or sound problems can only therefore be reasonable to cater for the average person and conditions. The approach to solving a noise problem can be three-fold:

1. Reduce the noise emitted at the source by such devices as mufflers and mounting machinery on resilient pads.
2. Provide a reasonable degree of sound insulation to reduce the amount of sound transmitted.
3. Isolate the source and the receiver.

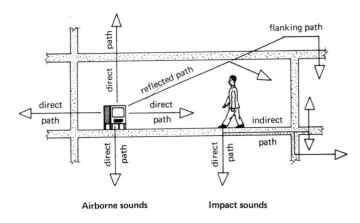

Airborne sounds   Impact sounds

Internal noise sources

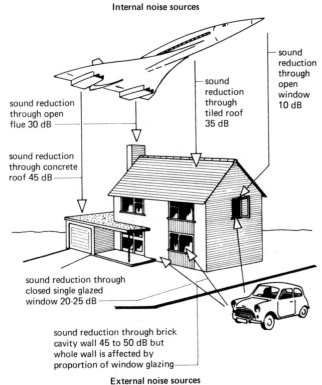

External noise sources

Fig. VI.1 Sources of noise within and around buildings

SOUND INSULATION

The most effective barrier to the passage of sound is a material of high mass. With modern materials and methods this form of construction is both impracticable and uneconomic. Unfortunately modern living with its methods of transportation and entertainment generates a considerable volume of noise and therefore some degree of sound insulation in most buildings is desirable.

BUILDING REGULATIONS

Part E of the Building Regulations deals with the sound insulation of dwellings and is intended to limit the amount of sound transmitted between dwellings and between rooms such as machinery rooms and adjacent dwellings.

Building Regulations E1, E2 and E3 dealing with airborne and impact sounds to walls and floors in the context of dwellings does not give numerical values for the sound insulation which is to be achieved. The requirements merely state that the wall or floor under consideration shall have reasonable resistance to airborne or impact sound.

The supporting document to these regulations, Approved Document E, gives various specifications and construction details which will satisfy the requirements of Part E of Schedule 1 to the Building Regulations. The Approved Document is divided into three sections, the first of which deals with walls and pays particular attention to the junctions of sound-resisting walls with floors, roofs and external walls and also to the problem of flanking sound. Flanking sound is the indirect transmission of sound around the end of a sound-resisting wall by passing through or around the flanking wall (see Fig. VI.2).

The second section of the Approved Document deals in a similar manner with the specifications and constructions for sound-resisting floors in the context of airborne or airborne and impact sound (see Fig. V.3). This section also gives guidance as to appropriate constructions for floor junctions and the penetration of pipes through floors.

Section 3 of the Approved Document consists mainly of tests which can be applied to show whether an existing wall or floor will give an acceptable performance in the context of sound transmission. A table giving acceptable sound transmission values for existing wall and floor construction is included in the Document.

External noise

Another aspect of sound insulation which, although not covered by Building Regulations, requires consideration is insulation against external noise. The main barrier to external noise is

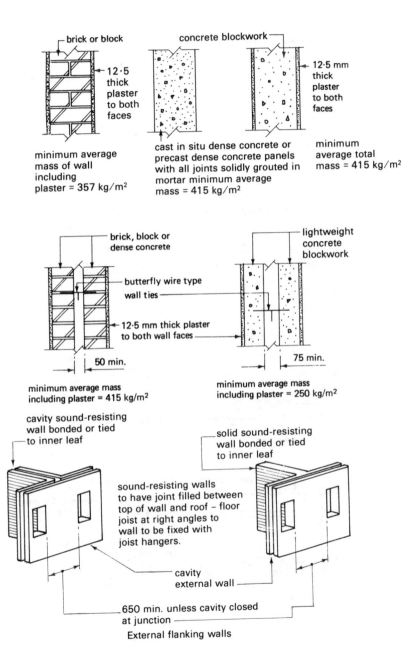

brick or block

12·5 thick plaster to both faces

minimum average mass of wall including plaster = 357 kg/m²

concrete blockwork

cast in situ dense concrete or precast dense concrete panels with all joints solidly grouted in mortar minimum average mass = 415 kg/m²

12·5 mm thick plaster to both faces

minimum average total mass = 415 kg/m²

brick, block or dense concrete

butterfly wire type wall ties

12·5 mm thick plaster to both wall faces

50 min.

minimum average mass including plaster = 415 kg/m²

lightweight concrete blockwork

75 min.

minimum average mass including plaster = 250 kg/m²

cavity sound-resisting wall bonded or tied to inner leaf

solid sound-resisting wall bonded or tied to inner leaf

sound-resisting walls to have joint filled between top of wall and roof – floor joist at right angles to wall to be fixed with joist hangers.

cavity external wall

650 min. unless cavity closed at junction

External flanking walls

Fig. VI.2 External flanking walls

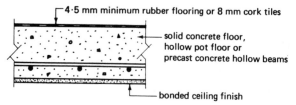

4·5 mm minimum rubber flooring or 8 mm cork tiles

solid concrete floor,
hollow pot floor or
precast concrete hollow beams

bonded ceiling finish

Concrete floors ~ minimum mass of floor = 365 kg/m²

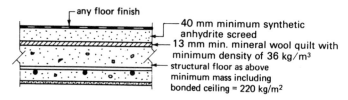

any floor finish

40 mm minimum synthetic
anhydrite screed
13 mm min. mineral wool quilt with
minimum density of 36 kg/m³
structural floor as above
minimum mass including
bonded ceiling = 220 kg/m²

Floating screed

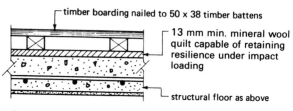

timber boarding nailed to 50 x 38 timber battens

13 mm min. mineral wool
quilt capable of retaining
resilience under impact
loading

structural floor as above

Floating timber raft

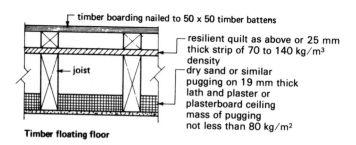

timber boarding nailed to 50 x 50 timber battens

resilient quilt as above or 25 mm
thick strip of 70 to 140 kg/m³
density
dry sand or similar
pugging on 19 mm thick
lath and plaster or
plasterboard ceiling
mass of pugging
not less than 80 kg/m²

joist

Timber floating floor

Fig. VI.3 Sound insulation – floors

192

provided by the shell or envelope of the building, the three main factors being:

1. The mass of the enclosing structure.
2. The continuity of the structure.
3. Isolation by double leaf construction when using lightweight materials.

Generally the main problem for the insulation against external noise is the windows, particularly if these can be opened for ventilation purposes. Windows cannot provide the dual function of insulation against noise and ventilation, since the admission of air will also admit noise. Any type of window when opened will give a sound reduction of about 10 decibels as opposed to the 45–50 decibel reduction of the traditional cavity wall. A closed window containing single glazing will give a reduction of about 20 decibels or approximately half that of the surrounding wall. It is obvious that the window to wall ratio will affect the overall sound reduction of the enclosing structure.

Double glazing can greatly improve the sound insulation properties of windows provided the following points are observed:

1. Sound insulation increases with the distance between the glazed units; for a reduction of 40 decibels the airspace should be 150–200 mm wide.
2. If the double windows are capable of being opened they should be weather-stripped.
3. Sound insulation increases with glass thickness particularly if the windows are fixed; this may mean the use of special ventilators having specific performances for ventilation and acoustics.
4. Double glazing designed to improve the thermal properties of a window have no real value for sound insulation.

Roofs of traditional construction and of reinforced concrete generally give an acceptable level of sound insulation, but the inclusion of rooflights can affect the resistance provided by the general roof structure. Lightweight roofing such as corrugated asbestos will provide only a 15–20 decibel reduction but is generally acceptable on industrial buildings where noise is generated internally by the manufacturing processes. The inclusion of rooflights in this type of roof generally has no adverse effects since the sound insulation values of the rooflight materials are similar to those of the coverings.

Modern buildings can be designed to give reasonable sound insulation and consequent comfort to the occupiers but the improvement to existing properties can present problems. A useful source of information on the reduction of noise in existing buildings is the Department of Environment advisory leaflet 69 obtainable from Her Majesty's Stationery Office.

26
Thermal insulation

Thermal insulation may be defined as a barrier to the natural flow of heat from an area of high temperature to an area of low temperature. In buildings this flow is generally from the interior to the exterior. Heat is a form of energy consisting of the ceaseless movement of tiny particles of matter called molecules; if these particles are moving fast they collide frequently with one another and the substance becomes hot. Temperature is the measure of hotness and should not be confused with heat.

The transfer of heat can occur in three ways:

Conduction: vibrating molecules come into contact with adjoining molecules and set them vibrating faster and hence they become hotter; this process is carried on throughout the substance without appreciable displacement of the particles.

Convection: transmission of heat within a gas or fluid caused by the movement of particles which become less dense when heated and rise thus setting up a current or circulation.

Radiation: heat is considered to be transmitted by radiation when it passes from one point to another without raising the temperature of the medium through which it travels.

In a building all three methods of heat transfer can take place since the heat will be conducted through the fabric of the building and dissipated on the external surface by convection and/or radiation.

The traditional thick and solid building materials used in the past had a natural resistance to the passage of heat in large quantities, whereas the lighter and thinner materials used today generally have a low resistance to the transfer of heat. Therefore to maintain a comfortable and healthy internal temperature the external fabric of a building must be constructed of a combination of materials which will provide an adequate barrier to the transfer of heat.

Thermal insulation of buildings will give the following advantages:

1. Reduction in the rate of heat loss.
2. Lower capital costs for heating equipment.
3. Lower fuel costs.
4. Reduction in the risk of pattern staining.
5. Reduction of condensation and draughts thus improving the comfort of the occupants.

BUILDING REGULATIONS

Building Regulation L1 states that reasonable provision shall be made for the conservation of fuel and power in buildings. The requirements of this regulation are satisfied by limiting the total aggregate areas of rooflights and windows and by not exceeding the maximum U values for elements which are given in Approved Document L. The document gives the maximum thermal transmissions coefficient or 'U' value for various situations. 'U' values are expressed in $W/m^2 K$ which is the rate of heat transfer in watts (joules/sec) through $1 \ m^2$ of the structure for $1°$ Celsius difference between the air on the two sides of the structure.

To calculate a 'U' value the complete constructional detail must be known together with the following thermal properties of the materials and voids involved:

Thermal conductivity: called the 'k' value and is the measure of materials' ability to transmit heat and is expressed as the heat flow in watts per m^2 of surface area for a temperature difference of $1°$ Celsius per metre thickness and is expressed as W/mK. Values for 'k' can be obtained from tables published by the Institute of Heating and Ventilating Engineers or from manufacturer's catalogues.

Thermal resistivity: this is the ability of a material, regardless of size or thickness, to resist the passage of heat and is the reciprocal of the thermal conductivity and is expressed as $1/k$ or $m \ k/W$.

Thermal resistance: this is the product of thermal resistivity and the thickness in metres and is expressed as R or $m^2 \ k/W$.

Surface resistances: these are given as fixed values in the Building Regulations to provide a standardisation in the calculation of 'U' values.

To calculate the 'U' value of any combination of materials the total resistance of the structure is found and then the reciprocal of this figure will give the required value as shown in the following example:

Cavity wall of 103 mm brick outer skin, 50 mm wide cavity, 100 mm lightweight block inner skin and 16 mm two-coat plaster internally.

> brickwork: $k = 1.150$; $1/k = 0.87$; R_1 $0.87 \times 0.103 = 0.089$
> cavity: from IHVE guide $R_2 = 0.176$
> blockwork: $k = 0.245$; $1/k = 4.08$; $R_3 = 4.08 \times 0.100 = 0.408$
> plaster: $k = 0.461$; $1/k = 2.17$; $R_4 = 2.17 \times 0.016 = 0.035$
> surface resistances: fixed value $R_5 = 0.18$.

$$
\text{'}U\text{' value} = \frac{1}{R_1 + R_2 + R_3 + R_4 + R_5}
$$

$$
= \frac{1}{0.089 + 0.176 + 0.408 + 0.035 + 0.18}
$$

$$
= \frac{1}{0.726}
$$

$$
\backsimeq 1.4 \text{ W/m}^2\text{K}.
$$

The maximum 'U' value requirements of the Building Regulations are shown diagrammatically in Fig. VI.4. Approved Document L of the Building Regulations gives details of deemed to satisfy provisions regarding thermal insulation and specifies various materials and methods.

The thermal insulation of roofs can be carried out at rafter level, beneath the covering or at ceiling level. Generally rafter level insulation will use more material but can be applied as a combined roofing felt and insulation thus saving labour costs. The roof void will be warm and on sheltered sites it should not be necessary to protect the cistern and pipework against frost attack. Applying the insulation at ceiling level will reduce the amount of material required and will also reduce the heat loss into the roof space, but since the void is unheated the plumbing working housed in the roof space will need insulating against freezing temperatures (see Fig. VI.4). Cold roofs will need to be ventilated to comply with Building Regulation F2.

INSULATING MATERIALS

When selecting or specifying thermal insulation materials the following must be taken into consideration:

1. Resistance value of the material.
2. Need for a vapour barrier since insulating materials which become damp or wet, generally due to condensation, rapidly loose their

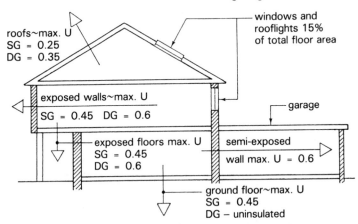

U values given in W/m²K ~ SG = single glazing
DG = total double glazing

roofs~max. U
SG = 0.25
DG = 0.35

windows and rooflights 15% of total floor area

exposed walls~max. U
SG = 0.45   DG = 0.6

garage

exposed floors max. U
SG = 0.45
DG = 0.6

semi-exposed wall max. U = 0.6

ground floor~max. U
SG = 0.45
DG – uninsulated

N.B. If calculated rate of heat loss through the solid parts of exposed elements is less than the maximum 'U' value trade-off calculations are permitted to ensure total heat loss of building is no more than permitted had the maximum 'U' values been used. Approved Document L gives worked examples of trade-off calculations using the following 'U' values for rooflights and windows – single glazing 5·7; double glazing 2·8; triple glazing 2·0

Maximum Permissible U Values for Dwellings

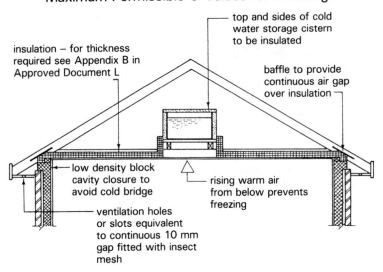

insulation – for thickness required see Appendix B in Approved Document L

top and sides of cold water storage cistern to be insulated

baffle to provide continuous air gap over insulation

low density block cavity closure to avoid cold bridge

rising warm air from below prevents freezing

ventilation holes or slots equivalent to continuous 10 mm gap fitted with insect mesh

Fig. VI.4 Building Regulation thermal insulation requirements

insulation properties; therefore if condensation is likely to occur a suitable vapour barrier should be included in the detail. Vapour barriers should always be located on the warm side of the construction.

3. Availability of material chosen.
4. Ease of fixing or including the material in the general construction.
5. Appearance if visible.
6. Cost in relation to the end result and ultimate savings on fuel and/or heating installation.
7. Fire risk — all wall and ceiling surfaces must comply with the requirements of Building Regulation B2 — restriction of spread of flame over surfaces of walls and ceilings.

Insulating materials are made from a wide variety of materials and are available in a number of forms:

Insulating concrete: basically a concrete of low density containing a large number of voids. This can be achieved by using lightweight aggregates such as clinker, foamed slag, expanded clay, sintered pulverised fuel ash, exfoliated vermiculite and expanded perlite, or alternatively an aerated concrete made by the introduction of air or gas into the mix. No fines concrete made by using lightweight or gravel aggregates between 20 and 10 mm size and omitting the fine aggregate is suitable for load bearing walls. Generally lightweight insulating concrete is used in the form of an *in situ* screed to a structural roof or as lightweight concrete blocks for walls.

Loose fills: materials which can be easily poured from a bag and levelled off between the joists with a shaped template. Materials include exfoliated vermiculite, fine glass fibrewool, mineral wool and cork granules. The depth required to give reasonable results is 25−35 mm; care should be taken to indicate, by paint or chalk marks on the sides of the joists, any electrical connections or junctions which have been covered over. Most loose fills are rot and vermin proof as well as being classed as non-combustible.

Boards: used mainly as dry linings to walls and ceilings either for self finish or direct decoration. Types include aluminium foil-backed plaster board, woodwool slabs, expanded polystyrene boards, asbestos insulating board and fibreboards. Insulating fibreboards should be conditioned on site before fixing to prevent buckling and distortion after fixing. A suitable method is to expose the boards on all sides so that the air has a free passage around the sheets for at least 24 hours before fixing. During this conditioning period the boards must not be allowed to become wet or damp.

Quilts: made from glass fibre or mineral wool bonded or stitched between outer paper coverings for easy handling. The quilts are supplied in rolls from 6.000–13.000 m long and cut to suit standard joist spacings. They are laid over the ceiling boards and can be obtained in two thicknesses, 25 mm thick for general use and 50 mm thick for use where a central heating system is installed.

Reflective insulation: used in both ceiling and wall insulation and consists of reinforced aluminium foil which should be used in conjunction with an unventilated cavity of at least 25 mm width.

Insulating plasters: factory produced pre-mixed plasters which have light-weight perlite and vermiculite expanded minerals as aggregates, and require only the addition of clean water before application. They are only one-third the weight of sanded plasters, have three times the thermal insulation value and are highly resistant to fire.

Foamed cavity fill: a method of improving the thermal insulation properties of an external cavity wall by filling the cavity wall with urea-formaldehyde resin foamed on site. The foam is formed using special apparatus by combining urea-formaldehyde resin, a hardener, a foaming agent and warm water. Careful control with the mixing and application is of paramount importance if a successful result is to be achieved; specialist contractors are normally employed. The foam can be introduced into the cavity by means of 25 mm bore holes spaced 1.000 m apart in all directions or by direct introduction into the open end of the cavity. The foam is a white cellular material containing approximately 99% by volume of air with open cells. The foam is considered to be impermeable and therefore unless fissures or cracks have occurred during the application it will not constitute a bridge of the cavity in the practical sense. The insertion of foam must comply with Building Regulation D1. In most cases the foam, upon setting, shrinks away from the inner face of the outer leaf enabling any water penetrating the outer leaf to run down the inside face of the external skin.

The most effective method of improving thermal comfort conditions within a building is to ensure that the inside surface is at a reasonably high temperature and this is best achieved by fixing insulating materials in this position.

Thermal insulation for buildings other than dwellings are covered by separate Building Regulations under Part L in which certain trade off calculations are permitted. The problems, materials and methods involved with these types of buildings are usually studied during the two years advanced technology applicable to most building courses.

27
Drainage

Part VII
Services

Drainage is a system of pipes, generally underground, used to convey the discharge from roofs, paved areas and sanitary fittings to a point of discharge or treatment. The discharges of rainwater and foul water can be conveyed together in a single drain or in separate drains to the public sewers according to the local authorities' directions based on their treatment facilities.

The fundamentals of drainage layouts, pipe materials, jointing techniques, bedding methods and simple inspection chambers construction together with the main requirements of the Building Regulations have been covered in Part IV of Volume I. A reiteration of these basic principles is not considered necessary and therefore the emphasis of this study will be on amplification and consideration of further points such as sewer connections, testing and soakaways in the context of simple drainage.

The arrangement of any drainage scheme is governed by:
1. Internal layout of sanitary fittings.
2. External positions of rainwater pipes.
3. Relationship of one building to another.
4. Location of public sewers.
5. Topography of the area to be served.

Drainage systems must be designed within the limits of the terrain, so that the discharges can flow by gravity from the point of origin to the point of discharge. The pipe sizes and gradients must be selected to provide sufficient capacity to cater for maximum flows and at the same time give adequate self-cleansing velocities at minimum flows to prevent deposits. Economic and constructional factors control the depth to which drains can be laid and it may be necessary in flat areas to provide pumping stations to raise the discharge to a higher level.

Private sewers

A sewer can be defined as a means of conveying waste, soil or rainwater below the ground that has been collected from the drains and conveying it to the final disposal point. If the sewer is owned and maintained by the local authority it is generally called a public sewer whereas one owned by a single person or a group of people and maintained by them can be classed as a private sewer.

When planning the connection of houses to the main or public sewer one method is to consider each dwelling in isolation but important economies in design can be achieved by the use of a private sewer. A number of houses are connected to the single sewer which in turn is connected to the public sewer. Depending upon the number of houses connected to the private sewer and the distance from the public sewer the following savings are possible:

1. Total length of drain required.
2. Number of connections to public sewer.
3. Amount of openings in the roads.
4. Number of inspection chambers.

A comparative example is shown in Fig. VII.1.

CONNECTIONS TO SEWERS

It is generally recommended that all connections to sewers shall be made so that the incoming drain or private sewer is joined to the main sewer obliquely in the direction of flow and that the connection will remain watertight and satisfactory under all working conditions. Normally sewer connections are made by the local authority or under their direction and supervision.

The method of connection will depend upon a number of factors:

1. Relative sizes of sewer and connecting drain or private sewer.
2. Relative invert levels.
3. Position of nearest inspection chamber on the sewer run.
4. Whether the sewer is existing or being laid concurrently with the drains or private sewers.
5. Whether stopped or joinder junctions have been built into the existing sewer.
6. The shortest and most practicable route.

If the public sewer is of a small diameter, less than 225 mm, the practical method is to remove two or three pipes and replace with new

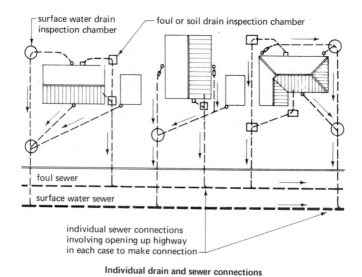

Individual drain and sewer connections

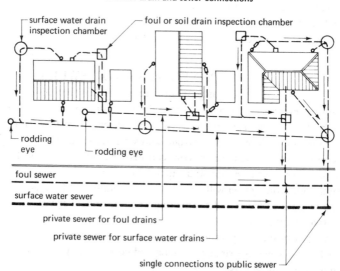

NB   generally 20 houses can be connected to a 100 mm dia.
     private sewer at a gradient of 1:70 and 100 houses
     can be connected to a 150 mm dia. private sewer
     laid to a fall of 1:150

Fig. VII.1 Example of a private sewer arrangement

pipes and an oblique junction to receive the incoming drain. If three pipes are removed it is usually possible to 'spring in' two new pipes and the oblique junction and joint in the usual manner, but if only two pipes are removed a collar connection will be necessary (see Fig. VII.2).

If new connections have been anticipated stopped junctions or joinder junctions may have been included in the sewer design. A stopped junction has a disc temporarily secured in the socket of the branch arm whereas the joinder has a cover cap as an integral part of the branch arm. In both cases careful removal of the disc or cap is essential to ensure that a clean undamaged socket is available to make the connection (see Fig. VII.2).

Connections to inspection chambers or manholes, whether new or existing, can take several forms depending mainly upon the differences in invert levels. If the invert levels of the sewer and incoming drain are similar the connection can be made in the conventional way using an oblique branch channel (see p. 234, Volume I). Where there is a difference in invert levels the following can be considered:

1. A ramp formed in the benching within the inspection chamber.
2. A backdrop manhole or inspection chamber.
3. Increase in the gradient of the branch drain.

The maximum difference between invert levels which can be success-fully overcome by the use of a ramp is a matter of conjecture. CP 2005 entitled 'Sewerage', gives a maximum difference of 1.800 m, whereas BS 8301, covering building drainage, gives 1.000 for external ramps at 45°. The generally accepted limit of invert level difference for the use of internal ramps is 700 mm, which approximates to the figure quoted in the first edition of CP 301. Typical constructional details are shown in Fig. VII.3.

Where the limit for ramps is exceeded a backdrop manhole construction can be considered. This consists of a vertical 'drop' pipe with access for both horizontal and vertical rodding. If the pipework is of clay or concrete the vertical pipe should be positioned as close to the outside face of the manhole as possible and encased in not less than 150 mm of mass concrete. Cast iron pipework is usually sited inside the chamber and fixed to the walls with holderbolts. Whichever material is used the basic principles are constant (see Fig. VII.3).

Changing the gradient of the incoming drain to bring its invert level in line with that of the sewer requires careful consideration and design. Although simple in conception the gradient must be such that a self-cleansing velocity is maintained and the requirements of Building Regulation Part H are not contravened.

Connections of small diameter drains to large diameter sewers can be made by any of the methods described above or by using a saddle connec-tion. A saddle is a short socketed pipe with a flange or saddle curved to

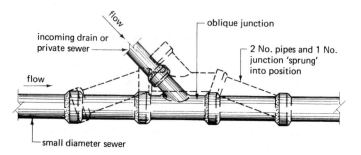

Removing 3 No. pipes and inserting oblique junction

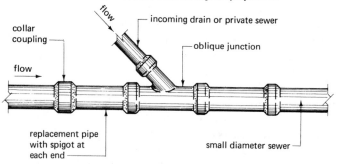

Removing 2 No. pipes and inserting oblique junction

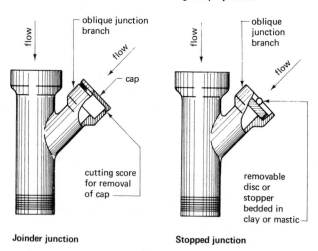

Joinder junction   Stopped junction

Fig. VII.2 Connections to small diameter sewers

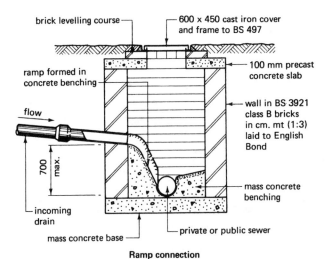

brick levelling course

600 x 450 cast iron cover
and frame to BS 497

ramp formed in
concrete benching

100 mm precast
concrete slab

flow

wall in BS 3921
class B bricks
in cm. mt (1:3)
laid to English
Bond

700 max.

mass concrete
benching

incoming
drain

mass concrete base

private or public sewer

Ramp connection

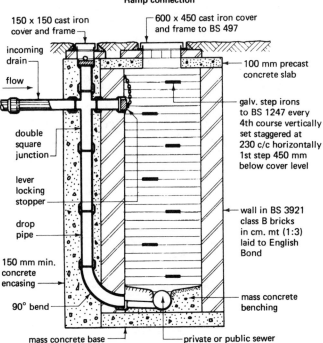

150 x 150 cast iron
cover and frame

600 x 450 cast iron cover
and frame to BS 497

incoming
drain

flow

100 mm precast
concrete slab

galv. step irons
to BS 1247 every
4th course vertically
set staggered at
230 c/c horizontally
1st step 450 mm
below cover level

double
square
junction

lever
locking
stopper

drop
pipe

wall in BS 3921
class B bricks
in cm. mt (1:3)
laid to English
Bond

150 mm min.
concrete
encasing

90° bend

mass concrete
benching

mass concrete base

private or public sewer

Backdrop connection

Fig. VII.3 Manhole and inspection chamber sewer connections

205

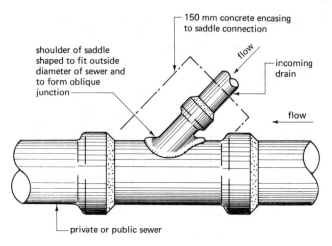

150 mm concrete encasing
to saddle connection

shoulder of saddle
shaped to fit outside
diameter of sewer and
to form oblique
junction

flow

incoming
drain

flow

private or public sewer

Connection arrangement

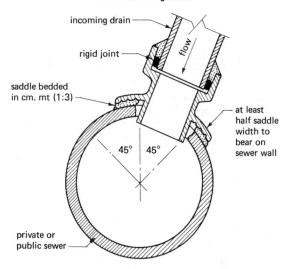

incoming drain

rigid joint

flow

saddle bedded
in cm. mt (1:3)

45° 45°

at least
half saddle
width to
bear on
sewer wall

private or
public sewer

NB saddle connection should be made in
the crown of the sewer pipe within 45°
on either side of the vertical axis

Typical section

Fig. VII.4 Saddle connections to sewers

suit the outer profile of the sewer pipe. To make the connection a hole must be cut in the upper part of the sewer to receive the saddle ensuring that little or no débris is allowed to fall into the sewer. A small pilot hole is usually cut first and this is enlarged to the required diameter by careful cutting and removing the débris outwards. The saddle connection is bedded on to the sewer pipe with a cement mortar and the whole connection surrounded with a minimum of 150 mm of mass concrete (see Fig. VII.4).

DRAIN TESTING

The Building Regulations 1985 make two references to the inspection and testing of drains. Regulation 14 requires that a person carrying out building work shall give the local authority notice in writing, or by such other means as they may agree, of the carrying out of any work of laying a drain or private sewer, including any necessary work of haunching or surrounding the drain and the backfilling of the trench. The required notice is to be given within not more than seven days after the work has been carried out.

Under Building Regulation 15 the local authority may make such tests of any drain or private sewer as may be necessary to establish whether it complies with the requirements of Part H (Drainage and waste disposal) of Schedule 1. Part H makes no direct reference to drain testing but the supporting Approved Document H recommends the water-testing of gravity drains and private sewers to establish watertightness and gives full details of the water-test requirements. The Approved Document also makes reference to the recommendations contained in BS 8301 which is the Code of Practice for building drainage.

The local authority will carry out drain testing after the backfilling of the drain trench has taken place, therefore it is in the contractor's interest to test drains and private sewers before the backfilling is carried out since the detection and repair of any failure discovered after backfilling can be time-consuming and costly.

Types of tests

There are three methods available for the testing of drains:

Water test: the usual method employed and is carried out by filling the drain run being tested with water under pressure and observing if there is any escape of water (see Fig. VII.5).

Smoke test: method used by some authorities by pumping smoke into the

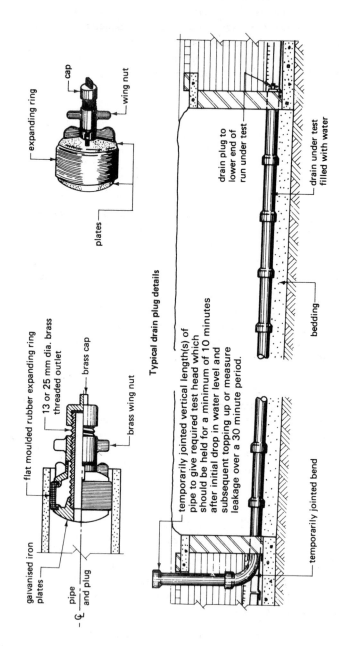

expanding ring

cap

wing nut

plates

Typical drain plug details

flat moulded rubber expanding ring

13 or 25 mm dia. brass threaded outlet

brass cap

brass wing nut

galvanised iron plates

Ҁ pipe and plug

temporarily jointed vertical length(s) of pipe to give required test head which should be held for a minimum of 10 minutes after initial drop in water level and subsequent topping up or measure leakage over a 30 minute period.

drain plug to lower end of run under test

drain under test filled with water

bedding

temporarily jointed bend

Fig. VII.5 Water testing of drains

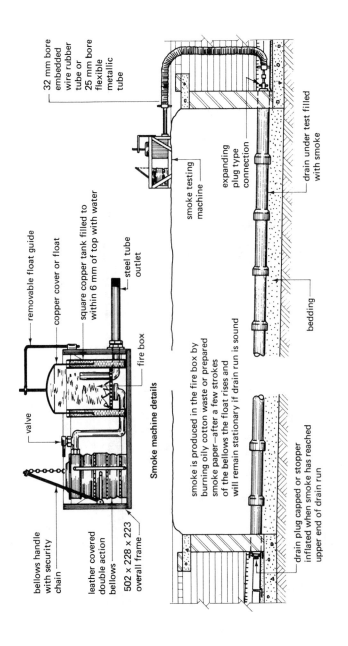

32 mm bore embedded wire rubber tube or 25 mm bore flexible metallic tube

removable float guide

copper cover or float

square copper tank filled to within 6 mm of top with water

steel tube outlet

fire box

valve

bellows handle with security chain

leather covered double action bellows

502 × 228 × 223 overall frame

Smoke machine details

smoke testing machine

expanding plug type connection

drain under test filled with smoke

bedding

smoke is produced in the fire box by burning oily cotton waste or prepared smoke paper—after a few strokes of the bellows the float rises and will remain stationary if drain run is sound

drain plug capped or stopper inflated when smoke has reached upper end of drain run

Fig. VII.6 Smoke testing of drains

209

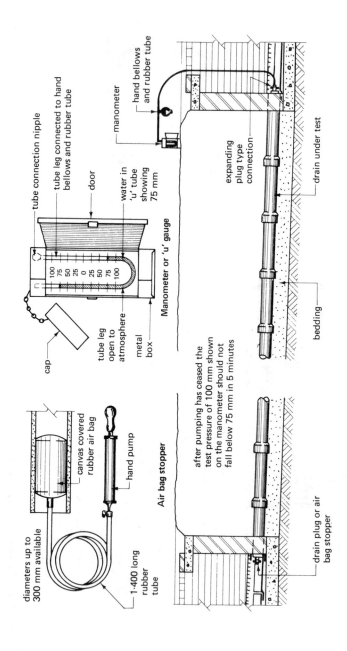

tube connection nipple

tube leg connected to hand bellows and rubber tube

door

water in 'u' tube showing 75 mm

manometer

hand bellows and rubber tube

cap

tube leg open to atmosphere

metal box

100 75 50 25 0 25 50 75 100

Manometer or 'u' gauge

expanding plug type connection

drain under test

bedding

diameters up to 300 mm available

canvas covered rubber air bag

hand pump

1·400 long rubber tube

Air bag stopper

after pumping has ceased the test pressure of 100 mm shown on the manometer should not fall below 75 mm in 5 minutes

drain plug or air bag stopper

Fig. VII.7 Air testing of drains

sealed drain run under test and observing any fall in pressure as indicated by the fall of the float on the smoke machine (see Fig. VII.6).

Air test: not a particularly conclusive test but it is sometimes used in special circumstances such as large diameter pipes where a large quantity of water would be required. If a failure is indicated by an air test the drain should be retested using the more reliable water test (see Fig. VII.7).

The illustrations of drain testing have been prepared on the assumption that the test is being carried out by the contractor before backfilling and haunching has taken place.

In general the testing of drains should be carried out between manholes; manholes should be tested separately and short branches of less than 6.000 m should be tested with the main drain to which they are connected; long branches would be tested in the same manner as a main drain.

SOAKAWAYS

A soakaway is a pit dug in permeable ground which receives the rainwater discharge from the roof and paved areas of a small building and is so constructed that the water collected can percolate into the surrounding subsoil. To function correctly and efficiently a soakaway must be designed taking into account the following factors:

1. Permeability or rate of dispersion of the subsoil.
2. Area to be drained.
3. Storage capacity required to accept sudden inflow of water such as that encountered during a storm.
4. Local authority requirements as to method of construction and siting in relation to buildings.
5. Depth of water table.

Before any soakaway is designed or constructed the local authority should be contacted to obtain permission and ascertain its specific requirements. Some authorities will permit the use of soakaways as an outfall to a subsoil drainage scheme or to receive the effluent from a small sewage treatment plant.

The rate at which water will percolate into the ground depends mainly on the permeability of the soil. Generally clay soils are unacceptable for soakaway construction, whereas sands and gravels are usually satisfactory. An indication of the permeability of a soil can be ascertained by observing the rate of percolation. A bore hole 150 mm in diameter should be drilled to a depth of 1.000 m. Water to a depth of 300 m is poured into the hole and the time taken for the water to disperse is noted. Several tests should

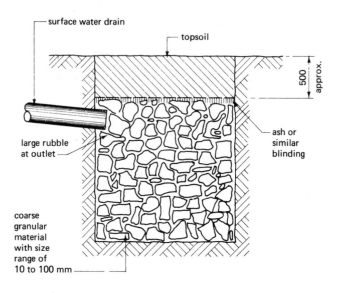

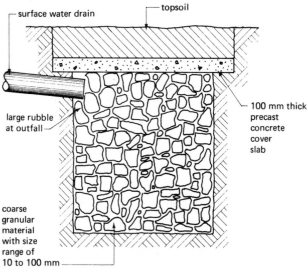

suitable coarse granular materials include broken
bricks, crushed sound rock and hard clinker

Fig. VII.8 Filled soakaways

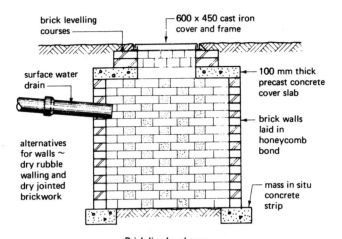

brick levelling courses

600 x 450 cast iron cover and frame

surface water drain

100 mm thick precast concrete cover slab

brick walls laid in honeycomb bond

alternatives for walls ~ dry rubble walling and dry jointed brickwork

mass in situ concrete strip

Brick-lined soakaway

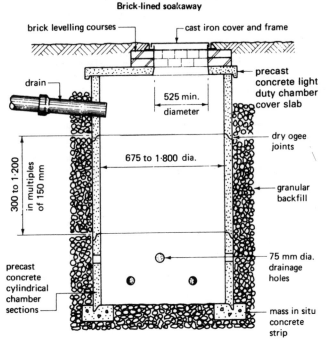

brick levelling courses

cast iron cover and frame

drain

525 min. diameter

precast concrete light duty chamber cover slab

dry ogee joints

675 to 1·800 dia.

granular backfill

300 to 1·200 in multiples of 150 mm

75 mm dia. drainage holes

precast concrete cylindrical chamber sections

mass in situ concrete strip

Precast concrete soakaway

Fig. VII.9 Lined soakaways

213

be made to obtain an average figure and the whole procedure repeated at 1.000 m stages until the proposed depth of the soakaway has been reached. A suitable diameter and effective depth for the soakaway can be obtained from a chart such as that illustrated in the Building Research Establishment Digest No. 151.

An alternative method is to calculate the volume of the soakaway by allowing for a storage capacity equal to one-third of the hourly rain falling on to the area to be drained. The rate of rainfall corresponding to a two-hour storm occurring on average not more than once in ten years is 0.015 m; therefore if the area to be drained is 150 m^2 the required capacity of the soakaway is:

$$150 \text{ m}^2 \times 0.015 \text{ m} = 2.25 \text{ m}^3.$$

Types of soakaways

A soakaway is constructed by excavating a pit of the appropriate size and either filling the void with selected coarse granular material or alternatively lining the sides of the excavation with brickwork or precast concrete rings (see Figs. VII.8 and VII.9).

Filled soakaways are usually employed only for small capacities, since it is difficult to estimate the storage capacity and the life of the soakaway may be limited by the silting up of the voids between the filling material. Lined soakaways are generally more efficient, have a longer life and if access is provided can be inspected and maintained at regular intervals.

Soakaways should be sited away from buildings so that foundations are unaffected by the percolation of water from the soakaway. The minimum 'safe' distance is often quoted as 3.000 m but local authority advice should always be sought. The number of soakaways required can only be determined by having the facts concerning total drain runs, areas to be drained and the rate of percolation for any particular site.

28
Domestic sanitary fittings and pipework

Sanitary fittings or appliances can be considered under two headings:

Soil fitments: those which are used to remove soil water and human excreta such as water closets and urinals.

Waste water fitments: those which are used to remove the waste water from washing and the preparation of food including appliances such as wash basins, baths, showers and sinks.

All sanitary appliances should be made from impervious materials, be quiet in operation, easy to clean and be of a convenient shape fixed at a suitable height. A number of materials are available for most domestic sanitary fittings including:

Vitreous china: a white clay body which is vitried and permanently fused with a vitreous glazed surface when fired at a very high temperature generally to the recommendations of BS 3402. Appliances made from this material are non-corrosive, hygienic and easily cleaned with a mild detergent or soap solution.

Glazed fireclay: consists of a porous ceramic body glazed in a similar manner to vitreous china; they are exceptionally strong and resistant to impact damage but will allow water penetration of the body if the protective glazing is damaged. Like vitreous china, these appliances are non-corrosive, hygienic and easily cleaned.

Vitreous enamel: this is a form of glass which can be melted and used to give a glazed protective coating over a steel or cast iron base. Used mainly for baths, sinks and draining boards, it produces a fitment which is lighter than those produced from a ceramic material, is hygienic, easy to clean and has a long life. The finish, however, can be chipped and is subject to staining especially from copper compounds from hot water systems.

Plastic materials: acrylic plastics, glass reinforced polyester resins and polypropylene sanitary fittings made from the above plastics require no protective coatings, are very strong, light in weight, chip resistant but generally cost more than ceramic or vitreous enamel products. Care must be taken with fitments made of acrylic plastics since these become soft when heated, therefore they should be used for cold water fitments or have thermostatically controlled mixing taps. Plastic appliances can be easily cleaned using warm soapy water and any dullness can be restored by polishing with ordinary domestic polishes.

Stainless steel: made from steel containing approximately 18% chromium and 8% nickle which gives the metal a natural resistance to corrosion. Stainless steel appliances are very durable and relatively light in weight; for domestic situations the main application is for sinks and draining boards. Two finishes are available: polished or 'mirror' finish and the 'satin' finish; the latter has a greater resistance to scratching.

The factors to be considered when selecting or specifying sanitary fitments can be enumerated thus:
1. Cost: outlay, fixing and maintenance.
2. Hygiene: inherent and ease of cleaning.
3. Appearance: size, colour and shape.
4. Function: suitability, speed of operation and reliability.
5. Weight: support required from wall and/or floor.
6. Design: ease with which it can be included into the general services installation.

WATER CLOSETS

Building Regulation G4 gives requirements for the receptacle and the flushing apparatus. Most water closets are made from a ceramic base to the requirements of BS 5503 with a horizontal outlet. The complete water closet arrangement consists of the pan, seat, flush pipe and flushing cistern. The cistern can be fixed as a high level, low level or closed coupled; the latter arrangement dispenses with the need for a flush pipe. A typical arrangement is shown in Fig. VII.10. The BS 5503 water closet is termed a wash down type and relies on the flush of water to wash the contents of the bowl round the trap and into the soil pipe. An alternative form is the siphonic water closet which is more efficient and quieter in operation but has a greater risk of blockage if misused. Two types are produced — the single trap and the double trap.

The single trap siphonic water closet has a restricted outlet which serves to retard the flow, when flushed, so that the bore of the outlet connected to the bowl becomes full and sets up a siphonic flushing action, completely emptying the contents of the bowl (see Fig. VII.10). With the double trap siphonic pan the air is drawn from the pocket between the two traps; when the flushing operation is started this causes atmospheric pressure to expel the complete contents of the bowl through both traps into the soil pipe (see Fig. VII.10).

The pan should be fixed to the floor with brass screws and bedded on a suitable compressible material; the connection to the soil pipe socket can be made with cement mortar or preferably a mastic to allow for any differential movement between the fitment and the structure. Connections to PVC soil pipes are usually made with compression rings. The flush pipe is invariably connected to the pan with a special plastic or rubber one-piece connector.

Flushing cisterns together with the flush pipes are usually constructed to the recommendations of BS 1125 and can be made from enamelled cast iron, enamelled pressed steel, ceramic ware or of plastic materials. Two basic types are produced, namely the bell or cone and the piston. The former is activated by pulling a chain which raises and lowers the bell or cone and in so doing raises the water level above the open end of the flush pipe thus setting up a siphonic action. These cisterns are efficient and durable but are noisy in operation (see Fig. VII.10). The piston type cistern is the one in general use and is activated by a lever or button. When activated the disc or flap valve piston is raised and with it the water level which commences the siphonage (see Fig. VII.10). The level of the water in the cistern is controlled by a ball valve and an overflow or warning pipe of a larger diameter than the inlet and is fitted to discharge so that it gives a visual warning, usually in an external position. The capacity of the cistern will be determined by local water board requirements, the most common being 9, 11.5 and 13.6 litres.

There is a wide range of designs, colours and patterns available for water closet suites but all can be classified as one of the types described above.

WASH BASINS

Wash basins for domestic work are usually made from a ceramic material but metal basins complying with BS 1329 are also available. A wide variety of shapes, sizes, types and colours are

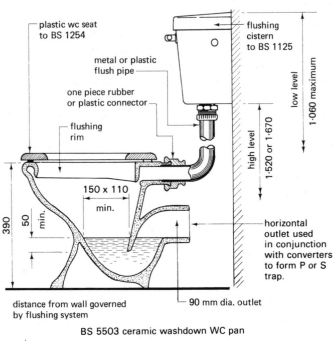

plastic wc seat
to BS 1254

metal or plastic
flush pipe

one piece rubber
or plastic connector

flushing
rim

flushing
cistern
to BS 1125

low level
1·060 maximum

high level
1·520 or 1·670

150 x 110
min.

390

50 min.

horizontal
outlet used
in conjunction
with converters
to form P or S
trap.

distance from wall governed
by flushing system

90 mm dia. outlet

BS 5503 ceramic washdown WC pan

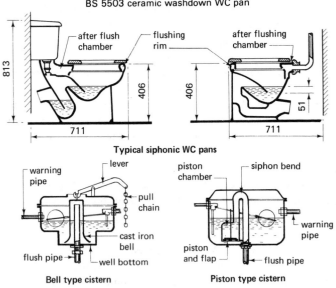

after flush
chamber

flushing
rim

after flushing
chamber

813

406

406

51

711

711

Typical siphonic WC pans

warning
pipe

lever

pull
chain

cast iron
bell

flush pipe

well bottom

piston
chamber

siphon bend

warning
pipe

piston
and flap

flush pipe

Bell type cistern

Piston type cistern

Fig. VII.10 WC pans and cisterns

available, the choice usually being one of personal preference. BS 1188 gives recommendations for ceramic wash basins and pedestals and specifies two basic sizes. 635 × 457 and 559 × 406. These basins are a one-piece fitment having an integral overflow, separate waste outlet and generally pillar taps (see Fig. VII.11).

Wash basins can be supported on wall-mounted cantilever brackets, leg supports or pedestals. The pedestals are made from identical material to the wash basin and are recessed at the back to receive the supply pipes to the taps and the waste pipe from the bowl. Although pedestals are designed to fully support the wash basin most manufacturers recommend that small wall mounted safety brackets are also used.

BATHS AND SHOWERS

Baths are available in a wide variety of designs and colours made either from porcelain-enamelled cast iron, vitreous enamelled sheet steel, 8 mm cast acrylic sheet or 3 mm cast acrylic sheet reinforced with a polyester resin/glass fibre laminate. Most bath designs today are rectangular in plan and made as flat bottomed as practicable with just sufficient fall to allow for gravity emptying and resealing of the trap. The British Standards for the materials quoted above recommend a co-ordinating plan size of 1 700 × 700 with a height within the range of n × 50 mm where n is any natural number including unity. Baths are supplied with holes for pillar taps or mixer fittings and for the waste outlets. Options include handgrips, built in soap and sponge recesses and overflow outlets. It is advisable to always specify overflow outlets as a precautionary measure to limit the water level and to minimise splashing; most overflow pipes are designed to connect with the waste trap beneath the bath (see Fig. VII.12). Support for baths is usually by adjustable feet for cast iron and steel and by a strong cradle for the acrylic baths. Panels of enamelled hardboard or moulded high impact polystyrene or glass fibre are available for enclosing the bath. These panels can be fixed by using stainless steel or aluminium angles or direct to a timber stud framework.

Shower sprays can be used in conjunction with a bath by fitting a rigid plastic shower screen or flexible curtain to one end of the bath. A separate shower fitment, however, is considered preferable. Such fitments require less space than the conventional bath, use less hot water and are considered to be more hygienic since the used water is being continuously discharged. A shower fitment consists of the shower tray with a waste outlet, the impervious cubicle and a door or curtain (see Fig. VII.12). Materials available are similar to those described for baths. The spray outlet is normally fixed to the wall and is connected to a mixing valve so that the water temperature can be controlled.

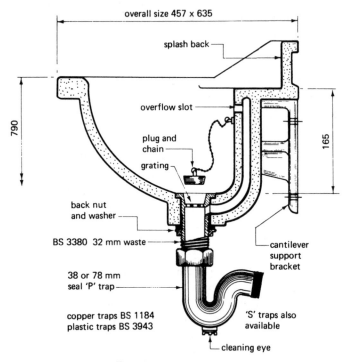

overall size 457 x 635

splash back

790

overflow slot

plug and chain

grating

165

back nut and washer

BS 3380 32 mm waste

38 or 78 mm seal 'P' trap

copper traps BS 1184
plastic traps BS 3943

cantilever support bracket

'S' traps also available

cleaning eye

Typical lavatory basin details

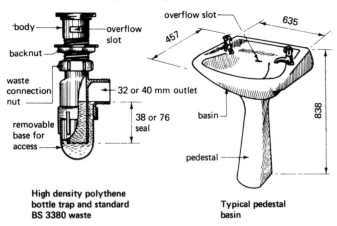

body

overflow slot

backnut

waste connection nut

removable base for access

32 or 40 mm outlet

38 or 76 seal

High density polythene bottle trap and standard BS 3380 waste

overflow slot

635

457

basin

838

pedestal

Typical pedestal basin

Fig. VII.11 Basins, traps and wastes

SINKS

Sinks are used mainly for the preparation of food, washing of dishes and clothes, and are usually positioned at the drinking water supply outlet. Their general design follows that described for basins except that they are larger in area and deeper. Any material considered suitable for sanitary appliance construction can be used. Designs range from the simple belfast sink with detachable draining boards of metal, plastic or timber to the combination units consisting of integral draining boards and twin bowls. Support can be wall-mounted cantilever brackets, framed legs or a purpose made cupboard unit; typical details are shown in Fig. VII.13.

The layout of domestic sanitary appliances is governed by size of fitments, personal preference, pipework system being used and the space available. Building Regulation G4 lays down specific requirements as to the interconnection of food storage and preparation rooms of sanitary accommodation which contains a water closet fitting.

PIPEWORK

Approved Document H sets out in detail the recommendations for soil pipes, waste pipes and ventilating pipes. These regulations govern such things as minimum diameters of soil pipes, material requirements, provision of adequate water seals by means of an integral trap or non-integral trap, the positioning of soil pipes on the inside of a building, overflow pipework and ventilating pipes. The only pipework which is permissible on the outside of the external wall is any waste pipe from a waste appliance situated at ground floor level providing such a pipe discharges into a suitable trap with a grating and the discharge is above the level of the water but below the level of the grating.

Three basic pipework systems are in use for domestic work, namely:
1. One-pipe system.
2. Two-pipe system.
3. Single stack system.

Whichever system is adopted the functions of quick, reliable and quiet removal of the discharges to the drains remains constant.

One-pipe system: consists of a single discharge pipe which conveys both soil and waste water directly to the drain. To ensure that water seals in the traps are not broken deep seals of 75 mm for waste pipes up to 65 mm diameter and 50 mm for pipes over 75 mm diameter are required. To allow for unrestricted layout of appliances most branch pipes will require an anti-siphon arrangement (see Fig. VII.14). The advantage of this system is

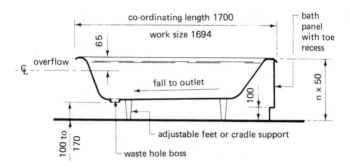

co-ordinating width = 700 work size = 697
co-ordinating sizes can be varied within range of
n x 100 mm where n = any natural number including unity

BS 1189 (cast iron) and BS 4305 (cast acrylic) baths

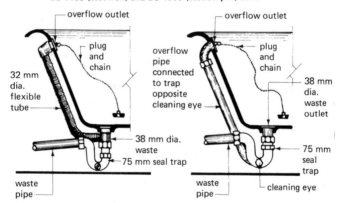

Alternative overflow connections

Typical Sizes:-
600 x 600 x 175 deep
760 x 760 x 175 deep
900 x 900 x 175 deep
also available in
enamelled steel
and perspex

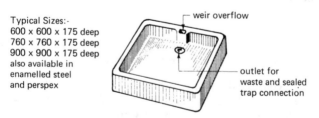

Typical fireclay shower tray

Fig. VII.12 Baths and shower trays

222

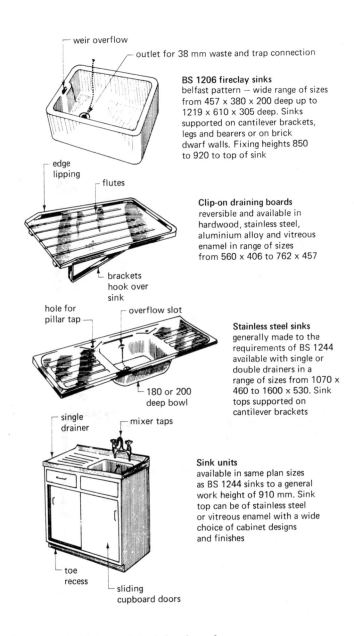

weir overflow

outlet for 38 mm waste and trap connection

BS 1206 fireclay sinks
belfast pattern – wide range of sizes
from 457 x 380 x 200 deep up to
1219 x 610 x 305 deep. Sinks
supported on cantilever brackets,
legs and bearers or on brick
dwarf walls. Fixing heights 850
to 920 to top of sink

edge
lipping

flutes

Clip-on draining boards
reversible and available in
hardwood, stainless steel,
aluminium alloy and vitreous
enamel in range of sizes
from 560 x 406 to 762 x 457

brackets
hook over
sink

hole for
pillar tap

overflow slot

Stainless steel sinks
generally made to the
requirements of BS 1244
available with single or
double drainers in a
range of sizes from 1070 x
460 to 1600 x 530. Sink
tops supported on
cantilever brackets

180 or 200
deep bowl

single
drainer

mixer taps

Sink units
available in same plan sizes
as BS 1244 sinks to a general
work height of 910 mm. Sink
top can be of stainless steel
or vitreous enamel with a wide
choice of cabinet designs
and finishes

toe
recess

sliding
cupboard doors

Fig. VII.13 Sinks and draining boards

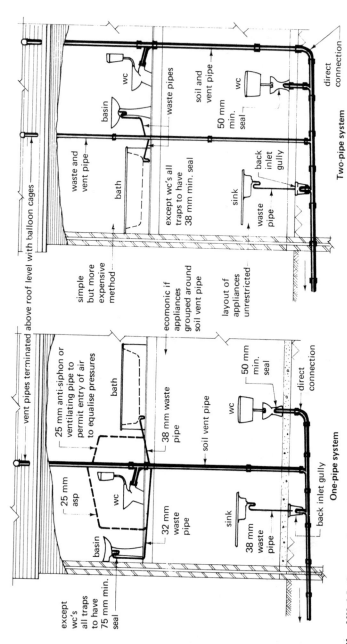

vent pipes terminated above roof level with balloon cages

except wc's all traps to have 75 mm min. seal

25 mm anti-siphon or ventilating pipe to permit entry of air to equalise pressures

25 mm asp

basin

wc

bath

32 mm waste pipe

38 mm waste pipe

soil vent pipe

sink

38 mm waste pipe

back inlet gully

wc

50 mm min. seal

direct connection

One-pipe system

simple but more expensive method

ecomonic if appliances grouped around soil vent pipe

layout of appliances unrestricted

wc

basin

waste pipes

waste and vent pipe

bath

except wc's all traps to have 38 mm min. seal

soil and vent pipe

50 mm min. seal

wc

back inlet gully

sink

waste pipe

direct connection

Two-pipe system

Fig. VII.14 Comparison of one-pipe and two-pipe systems

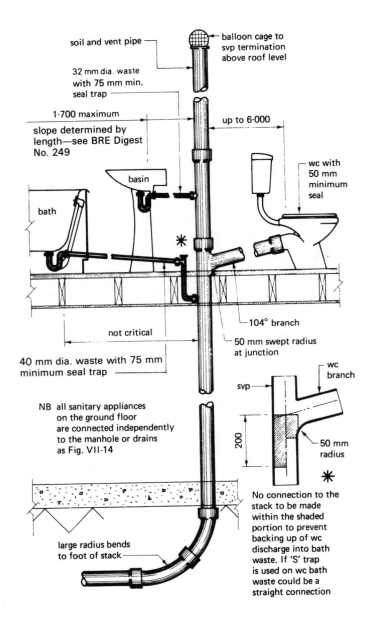

soil and vent pipe

balloon cage to svp termination above roof level

32 mm dia. waste with 75 mm min. seal trap

1·700 maximum

up to 6·000

slope determined by length—see BRE Digest No. 249

basin

wc with 50 mm minimum seal

bath

✱

104° branch

not critical

50 mm swept radius at junction

40 mm dia. waste with 75 mm minimum seal trap

wc branch

svp

NB all sanitary appliances on the ground floor are connected independently to the manhole or drains as Fig. VII-14

200

50 mm radius

✱

large radius bends to foot of stack

No connection to the stack to be made within the shaded portion to prevent backing up of wc discharge into bath waste. If 'S' trap is used on wc bath waste could be a straight connection

Fig. VII.15 Single stack system

the flexibility of appliance layout; the main disadvantage is cost; and generally the one-pipe system has been superseded by the more restricted but economic single stack system described later.

Two-pipe system: as its name implies, this system consists of two discharge pipes, one conveys soil discharges and the other all the waste discharges. It is a simple, reliable and costly system but has the advantages of complete flexibility in appliance layout and deep seal traps are not usually required. Like the one-pipe system, it has been largely superseded by the single stack system. A comparison of the one- and two-pipe systems is shown in Fig. VII.14.

Single stack system: this system was developed by the Building Research Establishment and is fully described in their Digest No. 249. It is a simplification of the one-pipe system by using deep seal traps, relying on venting by the discharge pipe and placing certain restrictions on basin waste pipes which has a higher risk to self siphonage than other appliances. A diagrammatic layout is shown on Fig. VII.15.

Materials which can be used for domestic pipework include galvanised steel (BS 3868) with socketed joints caulked with an asbestos material and cement in the form of a cord; cast iron (BS 416) with socketed joints sealed with hemp and caulked with run lead or a cold caulking compound and fixed like galvanised steel pipes by means of holderbolts to the support wall; pitch fibre (BS 2760) with push fit tapered joints or compression ring joints (it should be noted that not all local authorities will permit the use of pitch fibre for soil pipes fixed internally); UPVC (BS 4514), which can be jointed with a ring seal joint or by solvent welding; fixing to the support wall is by holderbolts or clips.

Most manufacturers of soil pipes, ventilating pipes and fittings produce special components for various plumbing arrangements and appliance layouts. These fittings have the water closet socket connections, bosses for branch waste connections and access plates for cleaning and maintenance arranged as one prefabricated assembly to ease site work and ensure reliable and efficient connections to the discharge or soil pipe.

29

Domestic electrical installations

A simple definition of the term electricity is not possible but it can be considered as a form of energy due to the free movement of tiny particles called electrons. If sufficient of these free or loose electrons move an electric current is produced in the material in which they are moving. Materials such as most metals and water which allow an electric current to flow readily are called conductors and are said to have a low resistance. Materials which resist the flow of an electric current such as rubber, glass and most plastics are called insulators.

For an electric current to flow there must be a complete path or circuit from the source of energy through a conductor back to the source. Any interruption of the path will stop the flow of electricity. The pressure which forces or pushes the current around the circuit is called the voltage. The rate at which the current flows is measured in amperes and the resistance offered by the circuit to passage of electrons is measured in ohms. A watt is the unit of power and is equal to the product of volts x amperes; similarly it can be shown that voltage is equal to the product of amperes x ohms.

Another effect of an electric current flowing through a conductor is that it will dissipate wasted energy in the form of heat according to the resistance of the conductor. If a wire of the correct resistance is chosen it will become very hot and this heating effect can be used in appliances such as cookers, irons and fires. The conductor in a filament bulb is a very thin wire of high resistance and becomes white hot thus giving out light as well as heat.

Most domestic premises receive a single phase supply of electricity from an area electricity board at a rating of 240 volts, and a frequency of 50 hertz. The area board's cable, from which the domestic supply is taken, consists of four lines, three lines each carrying a 240 volt supply and the fourth is the common return line or neutral which is connected to earth at the transformer or substation as a safety precaution should a fault occur on the electrical appliance. Each line or phase is tapped in turn together with the neutral to provide the single phase 240 V supply. Electricity is generated and supplied as an alternating current which means that the current flows first one way then the other; the direction change is so rapid that it is hardly discernible in such fittings as lights. The cycle of this reversal of flow is termed frequency.

The conductors used in domestic installations are called cables and consist of a conductor of low resistance such as copper or aluminium surrounded by an insulator of high resistance such as rubber or plastic. Cable sizes are known by the nominal cross sectional area of the conductor and up to 2.5 mm^2 are usually of one strand. Larger cables consist of a number of strands to give them flexibility. All cables are assigned a rating in amperes which is the maximum load the cable can carry without becoming overheated.

For domestic work wiring drawings are not usually required; instead the positions of outlets, switches and lighting points are shown by symbols on the plans. Specification of fittings, fixing heights and cables is given either in a schedule or in a written document (see Fig. VII.16).

RING CIRCUITS

Domestic buildings are wired using a ring circuit as opposed to the older method of having a separate fused sub-circuit to each socket outlet. Lighting circuits are carried out by using the 'loop in' method.

The supply or intake cable will enter the building through ducts and be terminated in the area board's fused sealing chamber which should be sited in a dry accessible position. From the sealing chamber the supply passes through the meter, which records the electricity consumed in units of kilowatt/hours, to the consumer unit which has a switch controlling the supply to the circuit fuses or miniature circuit breakers. These fuses or circuit breakers are a protection against excess current or overload of the circuit since should overloading occur the fuse or circuit breaker will isolate the circuit from the supply.

The number of fuseways or miniature circuit breakers contained in the consumer unit will depend upon the size of the building and the equipment to be installed. A separate ring circuit of 30 amp loading should be allowed for every 100 m^2 of floor area and as far as practicable the number

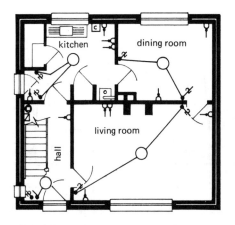

Ground floor plan

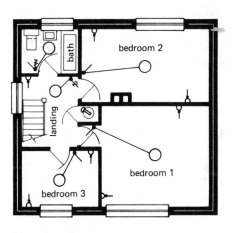

Upper floor plan

Symbols

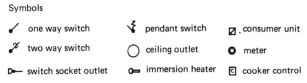

	one way switch		pendant switch		consumer unit
	two way switch		ceiling outlet		meter
	switch socket outlet		immersion heater		cooker control

Fig. VII.16 Typical domestic electrical layout

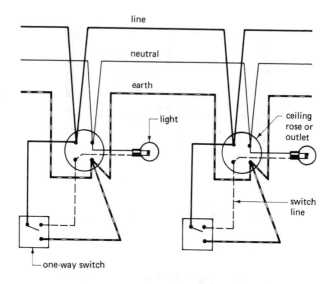

line

neutral

earth

light

ceiling
rose or
outlet

switch
line

one-way switch

'Loop-in' method of wiring using sheathed cable

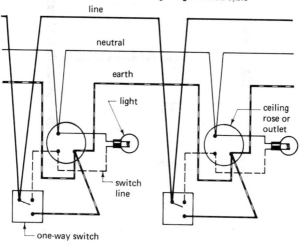

line

neutral

earth

light

ceiling
rose or
outlet

switch
line

one-way switch

Single core cable looped from switch to switch

Fig. VII.17 Typical lighting circuits

230

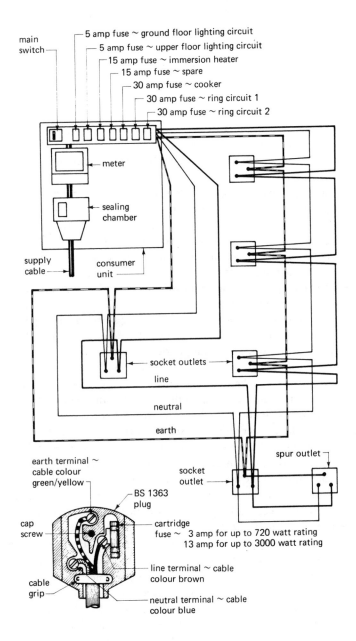

main switch

5 amp fuse ~ ground floor lighting circuit
5 amp fuse ~ upper floor lighting circuit
15 amp fuse ~ immersion heater
15 amp fuse ~ spare
30 amp fuse ~ cooker
30 amp fuse ~ ring circuit 1
30 amp fuse ~ ring circuit 2

meter

sealing chamber

supply cable

consumer unit

socket outlets

line

neutral

earth

spur outlet

socket outlet

earth terminal ~ cable colour green/yellow

BS 1363 plug

cap screw

cartridge fuse ~ 3 amp for up to 720 watt rating
13 amp for up to 3000 watt rating

line terminal ~ cable colour brown

cable grip

neutral terminal ~ cable colour blue

Fig. VII.18 Ring circuits and plug wiring

231

of outlets should be evenly distributed over the circuits. A typical domestic installation would have the following circuits from the consumer unit:

1. 5 amp: ground floor lighting up to ten fittings or a total load of 5 amps.
2. 5 amp: first floor lighting as above.
3. 30 amp: cooker circuit.
4. 30 amp: ring circuit 1.
5. 30 amp: ring circuit 2.
6. 15 amp: immersion heater.

The complete installation is earthed by connecting the metal consumer unit casing to the sheath of the supply cable or by connection to a separate earth electrode.

For lighting circuits using sheathed wiring a 1.0 mm^2 conductor is required and therefore a twin with earth cable is used. The 'loop-in' method of wiring is shown diagrammatically in Fig. VII.17. It is essential that lighting circuits are properly earthed since most domestic fittings and switches contain metal parts or fixings which could become live should a fault occur. Lighting circuits using a conduit installation with single core cables can be looped from switch to switch as shown in Fig. VII.17. Conduit installation consists of metal or plastic tubing together with various boxes for forming junctions and housing switches which gives a protected rewireable system. If steel-screwed conduit is used it will also serve as the earth leakage path but plastic conduit, being non-conductive, the circuit must have an insulated earth conductor throughout.

A ring circuit for socket outlets consists of a twin 2.5 mm^2 earthed cable starting from and returning to the consumer unit. The cables are looped into the outlet boxes making sure that the correct cable is connected to the correct terminals (see Fig. VII.18). The number of outlets is unlimited if the requirement of 1 ring circuit per 100 m^2 of floor area has been adopted. Spur outlets leading off the main ring circuit are permissible provided the limitation of not more than two outlet sockets on any one spur and not more than half the socket outlets on the circuit are on spurs is not exceeded. Socket outlets can be switched controlled and of single or double outlet; the double outlet is considered the best arrangement since it discourages the use of multiple adaptors. Fixed appliances such as wall heaters should be connected directly to a fused spur outlet to reduce the number of trailing leads. Moveable appliances such as irons, radios and standard lamps should have a fused plug for connection to the switched outlet, conforming to the requirements of BS 1363. The rating of the cartridge fuse should be in accordance with rating of the appliance. Appliances with a rating of not more than 720 watts should be protected

232

by a 3 amp fuse and appliances over this rating up to 3 000 watts should have a 13 amp fuse. As with the circuit, correct wiring of the plug is essential (see Fig. VII.18).

The number of outlets is not mandatory but the minimum numbers recommended for various types of accommodation are:

Kitchens: 3 plus cooker control unit with one outlet socket.
Living rooms: 3.
Dining rooms: 2.
Bedrooms: 2.
Halls: 1.
Landings: 1.
Garages: 1.
Stores and workshops: 1.

The outlets should be installed around the perimeter of the rooms in the most convenient and economic positions to give maximum coverage with minimum amount of trailing leads.

Cables sheathed with tough rubber or PVC can be run under suspended floors by drilling small holes on the neutral axis of the joists; where the cables are to be covered by wall finishes or floor screed they should be protected by either oval conduit or by means of small metal cover channels fixed to the wall or floor. Systems using mineral-insulated covered cables follow the same principles. This form of cable consists of single strands of copper or aluminium all encased in a sheath of the same metal which is densely packed with fine magnesium oxide insulation around the strands. This insulating material is unaffected by heat or age and is therefore very durable but it can absorb moisture. The sealing of the ends of this type of cable with special sealing 'pots' is therefore of paramount importance. Cables used in a conduit installation have adequate protection but it is generally necessary to chase the walls of the building to accommodate the conduit, outlet socket boxes and switch boxes below the wall finish level. Surface run conduit is normally secured to the backing by using screwed shaped clips called saddles.

The installation of electric circuits and electrical equipment are not covered by the Building Regulations but the minimum standard required by most authorities are those contained in the 'Regulations for Electrical Equipment of Buildings' issued by the Institution of Electrical Engineers.

30

Domestic gas installations

Gas is a combustible fuel which burns with a luminous flame; it is used mainly in domestic installations as a source of heat in appliances such as room heaters, cookers and water heaters. Gas can also be utilised to provide the power for washing machines and refrigerators; the use of gas as a means of artificial lighting has been superseded by electricity.

Gas is supplied by area boards under the general co-ordination and guidance of the Gas Council. The supply may be in the form of a manufactured gas known as 'town gas' or as a natural gas often referred to as 'North Sea Gas'. Town gas can be processed from coal, oil or imported natural gas and has a high hydrogen content of approximately 50% with a calorific value of about 18.6 MJ/m^3. Natural gas has no hydrogen content but has a very high percentage content of methane of approximately 95% with a calorific value of about 37.3 MJ/m^3. The pressure burning rate and amount of air required for correct combustion varies with the two forms of gas supply and therefore it is essential that the correct type of burner is fitted to the appliance.

The installation of a gas service to a domestic building is usually carried out by the local area board and can be considered in three distinct stages:

Main: this is the board's distribution system and works on the grid principle being laid and maintained by them. For identification purposes gas pipelines are colour coded over their length or in 150 mm wide bands with yellow ochre.

Service pipe: the connection pipe between the main and the consumer control which is positioned just before the governor and meter. In small domestic installations the service pipe diameter is 25−50 mm according to the number and type of appliances being installed. The pipe run should be as short as possible, at right angles to the main, laid at least 760 mm below the ground to avoid frost damage and be laid to a rise of 25 mm in 3 000 to allow for any condensate to drain back to a suitable condensate receiver.

Internal installation: commences at the consumer control and consists of a governor to stabilise the pressure and volume, the meter which records the volume of gas consumed, pipework to convey the gas supply to the appliances.

Pipework can be of mild steel, solid drawn copper pipes and flexible tubing of rubber or metallic pipe for use with portable appliances such as gas pokers. The size of installation pipes will depend upon such factors as gas consumption of appliances, frictional losses due to length of pipe runs and bends. Gas pipes are fixed by means of pipe hooks, clips or ring brackets at approximately 1.500 m centres. All pipes should be protected against condensation, dampness, freezing and corrosion; methods include painting with a red lead or bituminous paint or using plastic-coated copper pipework. Pipes which pass through walls are housed in a sleeve of non-corrodible material surrounded by packing of incombustible material such as asbestos to facilitate easy replacement and to accommodate small differential movements.

Gas appliances fall mainly into two groups:
1. Gas supply only: refrigerators, radiant heaters, gas pokers.
2. Gas supply plus other services: central heating units, water heaters, washing machines.

Gas refrigerators are silent, reliable, requiring little or no maintenance and cheap to run. They work on the absorption principle using ammonia dissolved in water. Free standing, built-in and table models are available; all require a firm level surface, a well ventilated position, clearance for opening of doors and access to the rear to make the fixed connection to the supply. Gas room heaters can be radiant heaters where the heat source is visible, convector heaters or in combination as a radiant-convector heater. Room heaters using gas have high efficiency, rapid heat build-up, easy to clean and low maintenance. Gas fires or room heaters will require a chimney and flue complying with the requirements of Building Regulations J2 and J3 and of the minimum flue size recommended in Approved Document J. Exceptions permitting discharge of gas appliances otherwise than into an open flue are given in the Approved Document. These include cookers in ventilated rooms, and appliances fitted in a bathroom, shower room or a private garage which must be designed and installed as balanced flue appliances. Room sealed or balanced flue heaters are appliances which have the heater body sealed from the room; the heater obtains air for combustion by an inlet connected directly with the external air and return the products of combustion to the external air by a separate outlet but usually to a common grill or guard. The termination of the flues are so designed that the external wind pressure effects are balanced, which obviates the need for a traditional flue.

Gas central heating consists of a boiler, connected to a flue complying with the Building Regulations, which provides the heating source for a fanned warm air system which is flexible in design and has a quick response or a circulated hot water system.

The installation of gas services, like those of electricity, are subjects which the student of building technology will study in detail regarding materials, methods, design and application during a two-year course of advanced work when all aspects of services are covered by a separate syllabus.

Bibliography

Relevant BS — British Standards Institution.
Relevant BSCP — British Standards Institution.
Building Regulations 1985 — HMSO.
Relevant B.R.E. Digests — HMSO.
Relevant advisory leaflets — DOE.
R. Barry. *The Construction of Buildings.* Crosby Lockwood & Sons Ltd.
Mitchells Building Construction Series. B. T. Batsford Ltd.
W. B. McKay. *Building Construction*, Vols. 1 to 4. Longman.
Specification. The Architectural Press.
A. J. Elder. *A. J. Guide to the Building Regulations.* The Architectural Press.
Relevant A. J. Handbooks. The Architectural Press.
R. Llewelyn Davies and D. J. Petty. *Building Elements.* The Architectural Press.
Cecil C. Handisyde. *Building Materials.* The Architectural Press.
W. Fisher Cassie and J. H. Napper. *Structure in Building.* The Architectural Press.
Handbook on Structural Steelwork — The British Constructional Steelwork Association Ltd. and The Constructional Steel Research and Development Organisation.
L. V. Leech. *Structural Steelwork for Students.* Butterworths.
Application of Mastic Asphalt — Mastic Asphalt Council and Employers Federation.
Gas Handbook for Architects and Builders — The Gas Council.
The Plumber's Handbook — Lead Development Association.

Copper Roofing — Copper Development Association.

Leslie Wolley. *Drainage Details*. Northwood Publications.

'Sanitation Details by Aquarius', *Building Trades Journal*. Northwood Publications.

Relevant manufacturers' catalogues contained in the Barbour Index and Building Products Index Libraries.

Index

239